STEM TRACTION

A STUDENT'S GUIDE TO ENGINEERING CAREERS

STEM TRACTION

ISBN: 979-8-9895436-1-8

Editor: Crystal S. Wright

10 9 8 7 6 5 4 3 2 1
Printed in the United States

Priceless
PUBLISHING

Priceless Publishing®
pricelesspublishing

DEDICATION

To my incredible husband Anderson —
thank you for always encouraging me to believe bigger.

Special thanks to Aria, Arion and Akira —
your curiosity and willingness to absorb new
information will forever keep me on my toes.

I love you all, beyond the galaxies and back.

—with endless love Marcia.

CONTENTS

FOREWORD

The writing of **STEMTRACTION: A STUDENT'S GUIDE TO ENGINEERING CAREERS** could not have come at a better time. This book is based on a conceptual framework that keeps science, technology, engineering, and mathematics (STEM) at the forefront of education, expatiating how to effectively embed STEM applications in learning, and creating an enhanced student engagement in STEM education.

As a Professor of Engineering at Texas Southern University in Houston, Texas, I have been engaged in various activities within the engineering profession for over 40 years, such as designing, researching, teaching, and even management of engineering projects. I have seen first-hand how Engineering graduates, and indeed STEM graduates in general, are impacting the world, unlike any other educational discipline. The majority of today's growth and advances in human endeavors had some element of engineering involved in their conception, resulting in long, fulfilling, and healthy lives for people.

The author writes in an exceptionally clear and understandable style that makes this book easy to read, even at the end of a busy day. The book is packed with special features, which give one a sense of what the field of Engineering requires. Succinct summaries of key, practical insights provided, are based on Marcia's in-depth and in-breadth experience. All information is supported by loads of references, so you can easily explore as deeply as you choose. The book takes a broad-ranging approach that enables readers to grasp the best insights from a wide variety of Engineering disciplines.

And presently — there has never been a stronger need for a book that lays out the foundations of good teaching at university levels in the STEM disciplines. Worldwide, STEM jobs are exploding at far higher rates than many other types of jobs, yet not enough candidates for these jobs are graduating from STEM programs. Only a small percentage of high school seniors are interested in pursuing STEM careers. A number of those students fall by the wayside as they bump against the challenges of STEM studies. But as Marcia lays out in this remarkably engaging book, there are ways to help improve students' desire and ability to master tough material.

The book emphasizes the need for exposure of school children starting at least in middle school, and preferably earlier, to engineering concepts and applications. Students are then better able to make informed choices about studying engineering and other technical fields. This book has indeed strengthened this notion and will help to open up important career opportunities for students.

It takes a learner-centered approach which will do much to stimulate student success. It contains up-to-date practical information about Engineering careers which will allow students and parents to make informed decisions about the STEM disciplines. Indeed, the understanding of STEM careers will greatly be enriched through the reading of this extraordinary book.

DAVID OLOWOKERE, PHD, PE

PROFESSOR OF ENGINEERING,
TEXAS SOUTHERN UNIVERSITY, HOUSTON, TX

INTRODUCTION

Welcome to **STEMtraction**! This book was born out of my passion for preparing students to enter the workforce. Teaching in a Historically Black University (HBCU) gives me the privilege to help underrepresented minorities excel in the field of engineering. I get to teach them the real-world applications of STEM and prepare them to overcome the negative social and cultural barriers that exist in this field. This book allows me to continue this work outside the classroom or office.

And if that was not enough, I am a mom to a curious little girl, Aria. It is such a joy to see her show interest in engineering too. From a young age, she has continuously explored the solar system and asked lots of questions which grew her interest in Aerospace Engineering. I began to search for age-appropriate resources to encourage her engineering interests but found none. So, I decided that it was up to me to create for Aria the tools that she sought. I also realized that my child's need is not unique. The sad reality is children are only interested in the areas that they are exposed to. The lack of diversity in STEM careers is not just about black women, as there is underrepresentation (and sometimes a lack of representation) of other racial and ethnic groups in STEM fields.

For the last fifteen years, I have worked proactively to provide a solution to this problem by working with pre-college students to add diversity to engineering. Data shows that there is an urgent need for youth to learn about STEM at an early age. This will expose them to the possibilities available to them, allowing them to explore and see what career options pique their interest from an early age.

STEMtraction uses STEM education as a paradigm shift in early learning, keeping the curiosity and genius streak of children intact through the growing years. **STEMtraction** features a learner-centered approach which will do much to stimulate student success. My goal is to make students more aware of the broad horizons of career choices available to them through STEM.

If you have a child that is not sure what they want to be when they grow up, this book may just be where they need to start. If your child is looking to explore career options or if you are certain that your child should be on the pre-engineering track, then **STEMtraction** is a great place to start.

Beyond this book, I have created STEMtraction, a premier Pre-College Engineering Educational organization aimed at providing enrichment opportunities in Science, Technology, Engineering, and Mathematics through hands-on learning, career exploration, engineering design, and financial literacy. STEMTraction was born out of my passion to bring engineering careers to students as early as the 5th grade through STEM exposure, by providing them with the skills and attributes needed to prepare them for college and beyond.

Head on over to **www.stemtraction.com** or ***@STEMtraction*** on Instagram to get STEM resources and more information on how to build the skills needed for college success and growth in the STEM industry.

Marcia

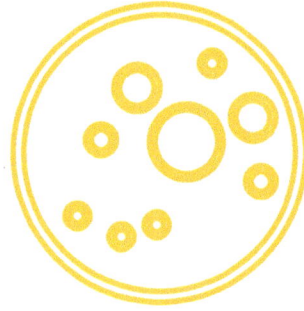

PART 1

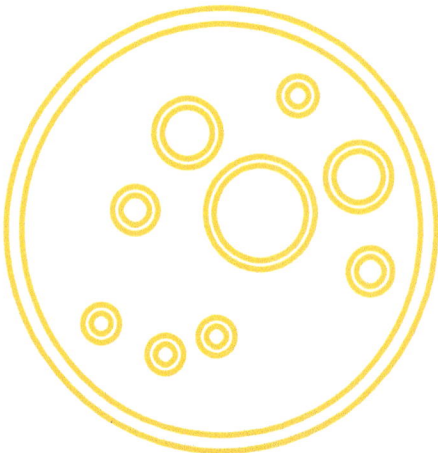

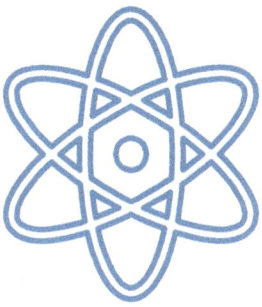

WHAT IS ENGINEERING?

Engineering is the application of mathematics and science to solve real-life problems. While scientists or inventors come up with ideas, it is the engineers who are responsible for making these ideas a reality. Some of the tasks performed by engineers include building bridges, roads, tunnels, vehicles, machines, and buildings.

DID YOU KNOW?

It was an engineer who designed and built the hearing aid?

This subject is a part of the Science Technology Engineering and Mathematics (STEM) category of subjects and is very enviable and lucrative as a career option. The role of the modern-day engineer is multi-faceted. They not only have to design, build, and test products and systems, but they must also oversee their safety, social and financial implications.

Engineering is responsible for so many everyday tasks that it is hard to imagine life without being impacted by the field. From your morning coffee to the transport you take, engineering touches every aspect of our lives.

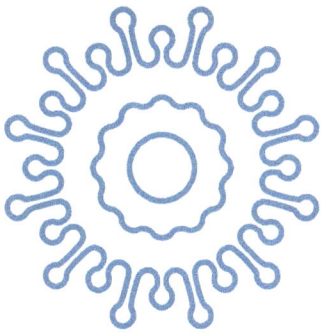

ENGINEERING FIELDS

There are many different fields of engineering. The four major ones are chemical, civil, mechanical, and electrical.

➡ **CHEMICAL ENGINEERING** is responsible for carrying out chemical processes on a commercial scale. Specialties include Manufacturing Engineer, Petroleum Engineer, Plastics Engineer, Process Engineer and Safety Engineer.

➡ **CIVIL ENGINEERING** is the field of engineering that covers the building of infrastructural works. The civil engineer is responsible for building works such as airports, drainage systems, roads and railways. It includes sub-disciplines such as surveying, environmental engineering, and structural engineering.

➡ **MECHANICAL ENGINEERING** refers to the application of principles and problem-solving techniques to the creation of machines. They design and manufacture mechanical systems. Mechanical engineers work to develop the machines that help in the manufacturing process.

→ **ELECTRICAL ENGINEERING** refers to the design, study, or manufacture of electronic and electrical circuits. For example: power, energy, weapons, transportation, engines, and compressors.

Engineering is a broad discipline that spans most of the industries in society and is responsible for the solution of a variety of problems. Engineering presents a challenging, fulfilling, practical approach to making life better for the general public. Engineers are responsible for some of the structural marvels that we see in the world today, such as the Burj Khalifa, Hoover Dam, and Three Gorges Dam!

DID YOU KNOW ?

The Department of Energy has been collecting atmospheric data for 25 years, operating 24 hours, 7 days a week?

HISTORY OF ENGINEERING

According to the Encyclopedia Britannica, the word *engineering* comes from the Latin root word *ingenerare* which means *to create*. The early English word *engine* meant *'to contrive'*.

Engineering was first considered a discipline in Mesopotamia (modern-day Iraq) in the 4th century BC when the wheel was invented. In Europe in the 14th century, an *engine'er* referred to a person responsible for constructing military engines such as catapults and other siege engines.

The discipline of engineering has its origins in the field of military equipment. Some of the first devices of war were floating bridges, assault towers, and catapults — all created by engineers! It gradually extended to civil society, to the building of roads, drains, bridges and buildings.

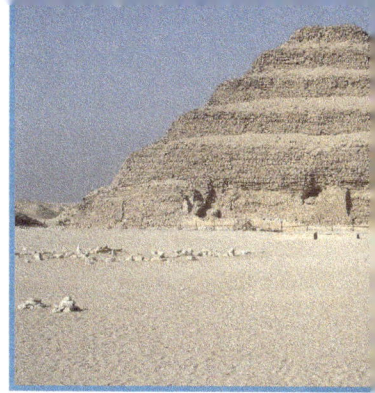

The first known engineer in the world was **Imhotep**. He was responsible for the construction of the Step Pyramid at Saqqarah, Egypt. There were other notable engineers who were responsible for the construction of road systems of the Greece and Persian empires, the lighthouse of Alexandria, Solomon's temple, and the Colosseum.

Vitruvius' De Architectura, published in the 1st century CE, is the first known thesis on engineering. The book covers ten volumes and includes chapters on building materials, construction methods, town planning, hydraulics, and measuring techniques. Engineering is an ancient discipline that has its origins in war and the construction of infrastructure of civil societies.

DID YOU KNOW?

Volt, the unit of electromotive force, was named after Alessandro Volta in 1881.

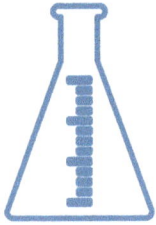

HIGH SCHOOL REQUIREMENTS FOR ENGINEERING

Since engineering falls in the STEM category, any student keen on pursuing the subject must be well-versed in Science and Mathematics. Some courses that a student should opt for at the secondary level are:

➡Robotics

➡Engineering or Design

➡Computer Science

➡Advance Placement (AP) Statistics

➡AP Physics *(Calculus-based is preferred)*

➡AP Calculus

MATHEMATICS

An applicant should be able to show that he or she has taken 4 years of mathematics in high school. Calculus and Statistics are very important subjects that students will delve deeply into in an engineering college program.

PHYSICS

Students need to opt for a calculus-based physics course instead of a non-calculus-based one. Failure to complete Algebra or regular-based Physics is a huge disadvantage for a high school student who wishes to pursue engineering. Some important topics covered in this area are Kinematics, Newton's Laws of Motion and Electromagnetism.

COMPUTER SCIENCE CLASSES

AP in Computer Science is also very useful as it helps students to gain an understanding of the realm of programming. Even if students are not desirous of pursuing computer science in the future, the course is a great starter course and an introductory course for engineers.

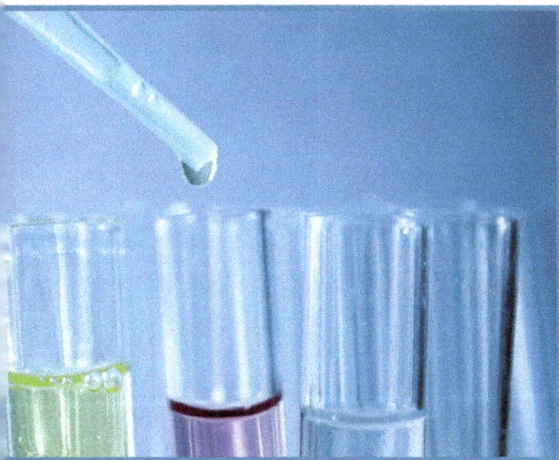

CHEMISTRY

Students should expect to take a course in general chemistry in college even if Chemical Engineering is not an option for them. AP Chemistry can be quite challenging, but it is a good option for increasing preparedness.

ECONOMICS

Expect to take at least one economics class in engineering program. Hence, having a basic knowledge of economics that will not be based on calculus is a good place to start. Other science subjects such as anatomy or astronomy are not required for a student looking to opt for engineering later. However, choose a challenging class as engineering colleges will look at the caliber of the classes you take and not the grade point average per se.

◀ ENGINEERING CAREER PATHS

Engineering is a prestigious and lucrative career option. However, the road to becoming an engineer can be bumpy. One needs to be suitably equipped to excel in this profession. On that note, let's look at the different levels of educational attainment one needs to achieve to fully qualify for this career. There are two routes that an individual pursuing a career in engineering can take: *degree-seeking* and *non-degree-seeking.*

DEGREE-SEEKING

UNDERGRADUATE COURSE (BACHELOR'S)

Potential engineers most often begin with a traditional undergraduate course. The undergraduate degree is typically 4 years long and involves courses in Biology, Physics, Mathematics, Chemistry, and Computer Science.

The final year of the program involves a Capstone course or a course that covers all the other programs studied over the four years. It is necessary to pass this Capstone course in order to graduate. You are eligible to pursue licensure after this stage.

POSTGRADUATE COURSE (MASTER'S)

To further specialize in the subject, students can choose to pursue a master's degree after their undergraduate studies are complete. Persons who have a desire or intention to increase their research knowledge should certainly consider pursuing a Master's. A master's degree will give you an advantage over your peers and give you an opportunity to increase your earning potential, work for large firms or further specialize in the field.

DOCTORAL COURSE (PHD/ENGD)

To further specialize, one can pursue a PhD or EngD. The main difference between these two doctoral level degrees is that the PhD is research-focused, while the EngD is practitioner-focused. Generally, people who pursue a PhD go on to focus on research or academia, while EngD is a tool to those who want to advance their skills further to become industry leaders.

Doctoral studies can further enhance career prospects for an individual and help them to attain their goals faster. One is likely to command a much higher salary and become eligible for high-level jobs and leadership roles with this level of education.

NON-DEGREE SEEKING

DIPLOMA

The other route for an engineering career after high school is to get a diploma. Although this is not as prestigious as an undergraduate degree, it can be pretty useful. A few states in the United States still allow individuals without an undergraduate degree to sit for the Fundamentals of Engineering (FE) exam if they have years of engineering experience under a professional engineer.

LICENSURE

To work as a qualified, professional engineer in the United States, one must first earn a four-year undergraduate degree from an ABET-accredited engineering program. The second step towards licensure is passing the Fundamentals of Engineering exam, also known as the FE. Success at this stage qualifies the individual as an *engineer in training* (EIT), or an *engineer intern* (EI).

Next, one needs work experience under a Professional Engineer (PE) for four years. After the four years are completed, one becomes eligible to sit the Professional Engineering licensing exam towards attaining the PE license. The PE license can open a multitude of doors professionally and give the holder desired credibility.

PART 2

ENGINEERING SPECIALTIES

There are numerous engineering specialties as we previously discussed. In this section we will discuss 15 of them.

ENGINEERING WONDERS

Engineers have made countless invaluable contributions to society. From the beginning of time, humankind has employed imagination and innovation to solve the problems of society and shape the world for the better. This has been accomplished through creativity, responsibility, and sound judgment.

As we continue to live in a global society, our problems are not limited to our own city or even our country. We are, rather, a global village affected collectively by the issues of environmental pollution, wastewater, drainage, greenhouse gasses, transportation, and medicine. As engineering includes a vast number of industries, you will come to learn that one invention lends progress to many different areas.

As society continues to advance technologically, engineering has profoundly shaped our modern lives. Integral to the world of engineering is one's ability to identify society's needs and respond to those needs using imagination, and technical and creative skills. Each specialty discussed in this section will highlight some wonders or practical applications as led by engineers within that specialty. This will provide some insight into the significant contribution of engineers.

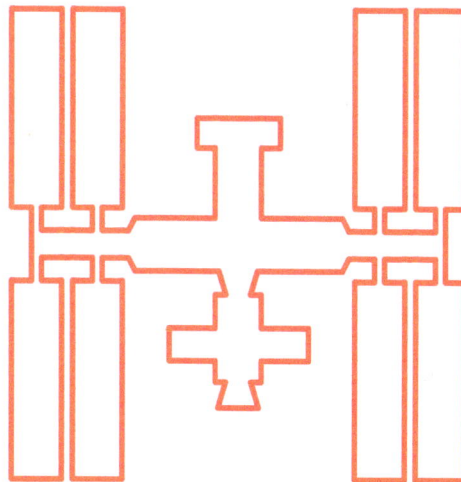

AEROSPACE ENGINEERING

DESCRIPTION OF FIELD

Field of engineering that deals with the design, testing, development, and production of spacecraft, aircraft, and related equipment. This field is further broken down into Aeronautical and Astronautical engineering. Aeronautical engineers focus on the theory, practice, and technology of flying within the Earth's atmosphere. Astronautical engineers focus on the science and technology of flying in outer space.

WHAT DO THEY DO?

Aerospace engineers have an in-depth understanding of aerodynamics, propulsion, vehicle dynamics, orbits, statics, and thermodynamics. Aerospace engineers design solutions to overcome the challenges of flying within the Earth's atmosphere or in outer space.

They develop the technologies that are used in airplanes and spaceships. They also design and manufacture airplanes, spacecraft, missiles, and satellites, as well as the components of these items.

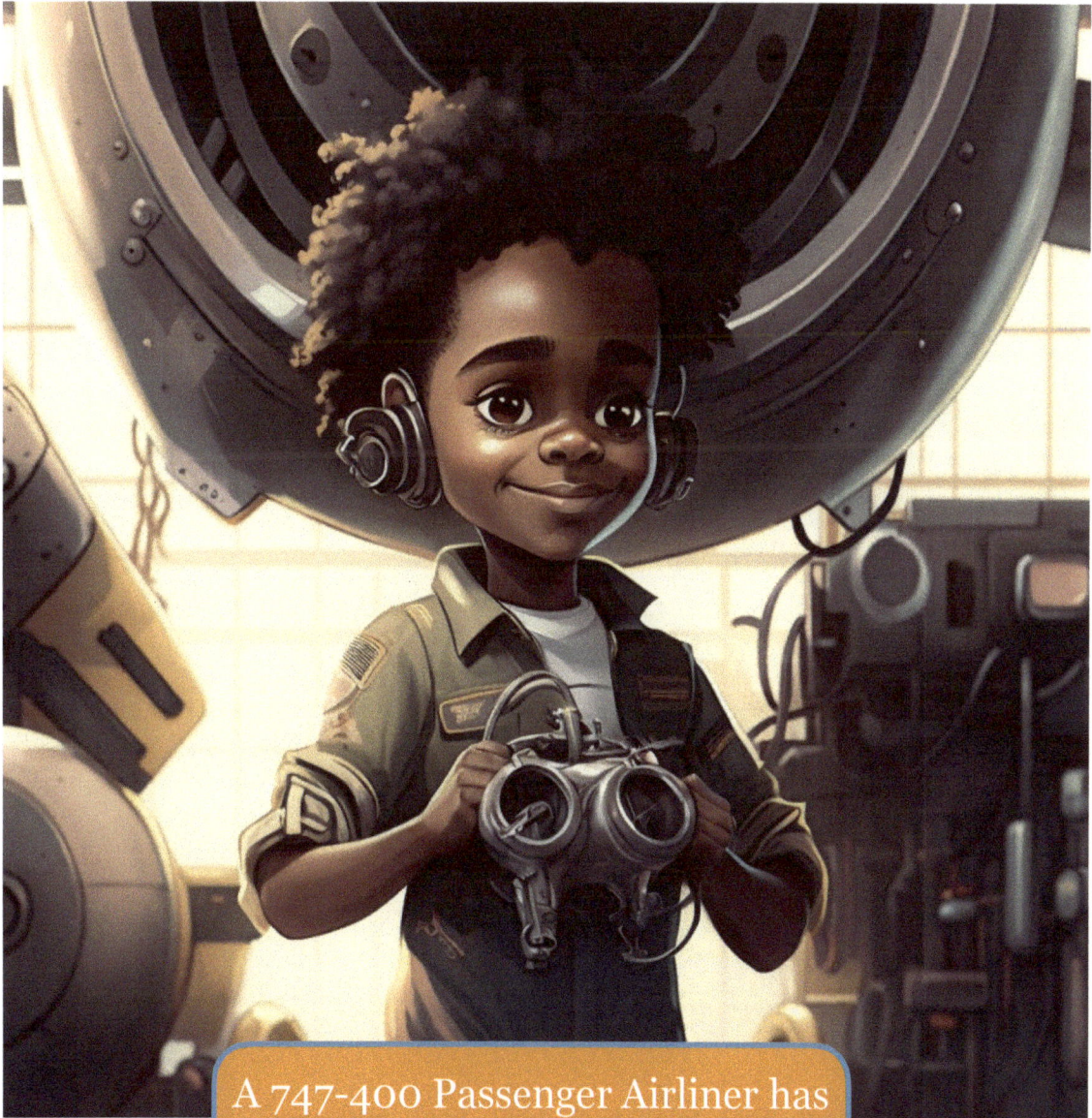

DID YOU KNOW?

A 747-400 Passenger Airliner has more than 6 million parts?

WHERE DO THEY WORK?

Careers include airline and parts manufacturers, aerospace and parts manufacturers, research facilities, and government agencies. Here is a list of employers of Aerospace Engineering and the industries they represent:

➡NASA

➡SpaceX

➡Boeing

➡Jet Propulsion Laboratory

NOTABLE HISTORICAL PERSONS IN THIS FIELD

✦ **Neil Armstrong** *(1930-2012)* was an American astronaut and aeronautical engineer. He was the first man to walk on the moon on June 20, 1969. The famous quote, *"That's one small step for man; one giant leap for mankind,"* was first said by him.

✦ **Joseph Samuel Dunning** *(1916-1991)* became the first black aeronautical engineer in the US in 1937.

ESTEEMED PERSONS OF MINORITY

✦ **Dr. Camille Wardrop Alleyne**, *(born in 1966)* in Trinidad, is an aerospace engineer.

✦ **Dr. Liselle Joseph** from Grenada is the first woman from Eastern Caribbean to earn a PhD in Aerospace Engineering.

DID YOU KNOW?

NASA Curiosity Rover has been on Mars for over 9 years?

WOMEN IN THE FIELD

✦ **Dr. Mae Carol Jemison** *(born in 1956)* is an engineer and former NASA astronaut. Jemison became the first African-American woman to travel to space in 1987!

✦ **Katherine Johnson** *(1918-2020)* was the first African-American woman to be employed by NASA. Johnson is also acknowledged as the first African American woman in the career of Aerospace Engineering. Her orbital mechanics' calculations were critical to the success of the first U.S. space missions.

✦ **Mary Jackson** *(1921-2005)* was the first African American female aerospace engineer to work at NASA.

CURRENT PERCENTAGE OF WOMEN & MINORITY EMPLOYED IN THE FIELD

➡ 12% FEMALE
➡ 71.9% WHITE
➡ 3.6% NON HISPANIC
➡ 5.91% BLACK

PROJECTED PERCENTAGE NEED FOR THE FUTURE.

➡ 2.71% 10 year projected growth

◀ FIRST PERSON IN SPACE

One of the most impressive feats of humanity was the launching of a Russian cosmonaut, Yuri Gagarin, into space on the 12th of April 1961. Yuri strapped himself into a tiny capsule called *Vostok 1*, and blasted into orbit. He almost completed one full orbit of the earth before landing safely back on earth.

He is recognized as the first human being to observe the earth from space. This feat was achieved during the decade-long space race between the then Soviet Union and the United States of America. The USA ultimately won the space race.

APOLLO PROGRAM

The human spaceflight program which succeeded in the preparation and transportation of the first humans to the moon and back is remarkable to this day. Apollo's launch vehicle known as the Saturn 5 rocket, was 10m in diameter, and 110m in height (about the height of the Statue of Liberty), with a thrust of 33400 kilo Newton (kN). It could send 43 tons to lunar orbit!

To put this in perspective, the most powerful rocket currently flying has a thrust of 22819 kN and can launch 16 tons to lunar orbit. There was a total of six missions to the surface of the moon starting on July 20, 1969. No one has returned since NASA's Apollo 17 mission in December 1972.

DID YOU KNOW ?

The output of the sun and stars can be used to produce energy? This type of energy is called *fusion*, and it is being explored in the United States as an energy source?

VOYAGER PROBES

The voyager program was developed to collect scientific data on Saturn and Jupiter, and *Voyager 2* went on to explore both Uranus and Neptune. The mission set by NASA was exploratory and many discoveries were made, including volcanoes and moons. The voyagers were sent on a trajectory that would take them out of the solar system. This trajectory calculated years earlier revealed that using the larger planets' gravity, a spaceship would be able to visit Neptune, Saturn, and Uranus. Such an occurrence happens every 176 years.

SPACEX

SpaceX's mission is to get humans living on other planets. Elon Musk, Chief Engineer of SpaceX, aims to create the next generation of fully reusable launch vehicles. SpaceX has managed to lower the costs of entering space by about 50% by modifying traditional fuel and engines, as well as recovering and re-using as much of the rocket and launch vehicle as possible. To date, SpaceX is the first company to take civilians and even a car into space!

In September 2008, SpaceX's Falcon became the first privately created liquid-fueled rocket to orbit the Earth. It is capable of carrying double payloads to as far as Mars and the moon! (Payload refers to any cargo housed on a rocket such as passengers, missiles, or satellites). SpaceX launched the first-ever all-civilian crew into orbit in 2021.

SOLAR IMPULSE

The Solar impulse is an aircraft that can fly using only the energy collected from solar panels. During the day, the solar panels on its wings charge its 4 lithium batteries, and at night the batteries power the engines. The world's largest airplane, the Airbus A380, has a wingspan of just under 80 meters and weighs over 600,000 pounds without passengers. The Solar Impulse 2 has a wingspan of 72 meters but weighs approximately 5000 pounds! That's about the same as the weight of an SUV.

In 2016, the Solar Impulse 2 completed the first around-the-world flight without using a single drop of fuel. The trip took 14 months, consisting of approximately 25,000 miles (23 days) of flight, totaling up to 550 hours in the sky. The aircraft contained 17, 248 photovoltaic cells, with a maximum power of 4 engines of 13.5 kilowatts.

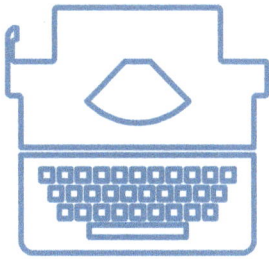

MECHANICAL ENGINEERING

DESCRIPTION OF FIELD

Concentrates on the study, design, testing, and manufacture of dynamic structures, such as tools, machinery, and engines. This specialization is one of the oldest and broadest engineering specialties, having emerged as a field in 19th-century Europe during the Industrial Revolution.

The American Society of Mechanical Engineers (ASME) was formed in 1880.

DID YOU KNOW?

In 1945 a blind mechanical engineer named Ralph Teetor invented the technology known as cruise control?

WHAT DO THEY DO?

Mechanical Engineers design elevators, escalators, electric generators, steam and gas turbines as well as refrigeration and air conditioning systems. Like many other engineers, they work closely with computers and are responsible for integrating sensors, controllers, and machinery. They use computer technology to create designs, run simulations, and test machines.

WHERE DO THEY WORK?

Careers include engine designers, heating and cooling systems engineer, design engineer, combustion engineers, and robotic engineers. Mechanical engineers can be hired in almost any field.

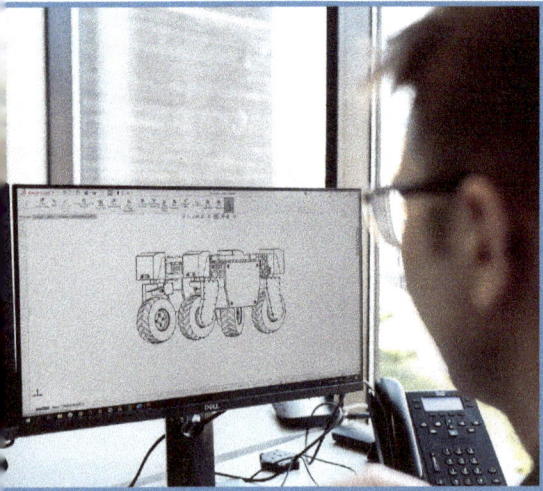

Here is a list of employers of mechanical engineers, and the industries they represent: NASA (aeronautics), Boeing (aerospace), Lockheed Martin Corporation (military equipment manufacturer), Microsoft (computer technology), Ford Motor Company (automobile manufacturer), and the United States Department of Energy (government).

Walt Disney Company hires mechanical engineers (called imagineers), to create entertainment experiences for patrons of the Walt Disney theme parks.

NOTABLE HISTORICAL PERSONS IN THIS FIELD

✦ **James Watt** *(1736 - 1819)* was a Scottish engineer referred to as the father of mechanical engineering, did not invent the steam engine but is credited with improving its design. Watt's remarkable invention employed a double-action engine and steam indicator. This was crucial to the development of the field of mechanical engineering. The steam engine led the way for other vehicles of transportation including steam locomotives and self-propelled boats. Soon, these inspired a growing network of railroad tracks and canals.

✦ **Nikola Tesla** *(1856 - 1943)* was well known for his contributions to the design of the modern alternating current electricity supply system. He also contributed to radio technologies and is often known as the father of radio.

✦ **George Stephenson**: This was a British inventor who created the first inner-city railway system. He is known as the father of railways. This, of course, reshaped the transportation industry."

ESTEEMED PERSONS OF MINORITY

✦ **Frederick Douglas Patterson** (first African American car manufacturer)

✦ **Granville Woods** (1856 – 1910) was the first inventor of African ancestry to be a mechanical and electrical engineer post the Civil War. He was a contemporary of George Westinghouse, Alexander Graham Bell, and Thomas Edison.

✦ **David Crosthwait** *(1898 – 1976)* was a Black electrical and mechanical engineer. He designed the heating system for Radio City Music Hall, and Rockefeller Center in New York City and received patents on 39 inventions related to the installation, design, service and testing of HVAC (heating, ventilation, air conditioning) power plants.

WOMEN IN THE FIELD

✦ **Bertha Lamme Feicht** *(1869 - 1943)* was the nation's first female mechanical engineer. She graduated from Ohio State University as the first American woman to achieve an engineering degree outside of civil engineering.

✦ **Lillian Moller Gilbreth** *(1878 - 1972)* became the first female engineer to earn a doctoral degree, though it was outside of the field. Moller is lauded as one of the female pioneers in the field of industrial-organizational psychology which she earned a PhD in. Her areas of expertise included workplace patterns and ergonomics.

These laid the foundation for what we now refer to as human factors engineering and ergonomic design. Moller's research and work led to wide scale improvements in workplace lighting, chairs and coffee breaks to reduce fatigue and increase productivity. She joined the American Society of Mechanical Engineers in 1926 as the first female.

✦ **Yvonne Young Clark** *(1929 - 2019)* was named the first woman to receive a Bachelor of Science degree in mechanical engineering in 1951 at Howard University. She has earned several awards, including the Distinguished Engineering Educator Award from the Society of Women Engineers in 1997.

CURRENT PERCENTAGE OF WOMEN & MINORITY EMPLOYED IN THE FIELD

➡ 91.3% male, about 9% female **(Census Bureau)**
➡ 74.4% White Non-Hispanic; 11.9% Asian Non-Hispanic; 3.8% Black **(Census Bureau)**

AVERAGE MALE SALARY

➡ $94,017

AVERAGE FEMALE SALARY

➡ $87,445

PROJECTED PERCENTAGE NEED FOR THE FUTURE

The projected percent change in employment from 2020 to 2030 is 7% (bls.gov)

The employment, or size, of this occupation in 2020, which is the base year of the 2020-30 employment projections was 299,200.

★ The projected numeric change in employment from 2020 to 2030 is 20,900 annually.

◀— EXAMPLES OF MECHANICAL ENGINEERING

NANO DRONES

Dronies or nano drones are self-flying quadcopters and are classified as the world's tiniest machine. They can dance coordinated without the use of external or on-board infrared sensors. Dronies are designed for indoor flying, using a 2.4 GHz transmitter with a signal strength that extends up to 30 meters. The recommended distance between the controller and the dancing dronies is 2-10 meters.

The coordinated flying and navigation abilities of dronies allow choreography based on 6-axis aerial acrobatics. A series of small drones exist around the world and the competition for the tiniest is in progress. A flying manufactured microchip which is the tiniest flying machine in the world — the size of a grain of sand — is being developed by Northwestern University scientists. It's made from biodegradable material and used for collecting data.

ROBO-FLY

The Robo-fly was designed by a Harvard scientist and created with electric muscles that enable the wings to be able to flap about 120 times per second! It is made from carbon fiber. After 12 years of research by the Robotics team, this microrobotics wonder came into being after the team solved two key technical challenges – a) building a sub-millimeter scale for precise and efficient measurements and b) creating artificial muscles for the fly. Application of this device can be used in environmental monitoring and also crop pollination.

LIEBHERR LTM 11200-9.1

Liebherr LTM 11200-9.1 is undoubtedly the longest and strongest mobile crane in the world. It can carry as many as 12 adult blue whales at once, with a maximum lifting capacity of 1300 tons. This 18-wheeler crane has an 8-part telescopic boom that can extend up to 328 feet. It has 9 axles and can drive at speeds of 75 miles per hour, with an extension capacity of 617 feet (50 stories)!

BAGGER 288

Bagger 288 or Excavator 288 is a bucket-wheel excavator and the world's largest land vehicle. It was manufactured by the German company Krupp and is used in the mining industry. It weighs approximately 13,000 tons! This excavator is taller than the Statue of Liberty, weighs more than the Eiffel Tower, and is bigger than the transporter that carries the NASA space shuttle. It can move as many as 40,000 workers or 10,000 dump trucks in a day, making this machine very economical.

NANO ENGINE

The University of Mainz in Germany has successfully completed the construction of the smallest working engine. For its size, the single charged atom will be used to heat and cool other nanomachines. In the future it will be used to power other nanomachines. The power generated is converted into a vibration of an atom which serves as a mechanical motion.

ELECTRICAL ENGINEERING

DESCRIPTION OF FIELD

Electrical Engineers develop, design, test and oversee the manufacturing process of electrical systems and equipment. These professionals use their specialized knowledge of electricity to solve human problems. Their work is necessary to support other industries, thus they have high job security and base pay rates. They are becoming more and more intertwined with computer engineering, as IT personnel work to automate electrical processes. The American Society of Electrical Engineers (ASCE) was formed in 1884.

WHAT DO THEY DO?

Electrical engineers work includes designing generators, telecommunication devices, motors, radar, navigation systems, and the electrical systems of automobiles and aircraft.

WHERE DO THEY WORK?

Electricity is required for all modern processes so Electrical Engineers work closely with engineers in other specialties. They can work as software developers. Careers and employers include IT, electricity manufacturing companies, broadcasting, and other engineering industries.

DID YOU KNOW ?

Alexander Graham Bell invented the telephone in 1876?

NOTABLE HISTORICAL PERSONS IN THIS FIELD

✦ **Miller Reese Hutchinson** *(1876-1944)* was an American electrical engineer. He was credited with inventing the first portable and electrical hearing aid in 1898.

✦ **Claude Shannon** *(1916 – 2001)* is widely considered as the father of information theory. He was a cryptographer, mathematician, and electrical engineer. Shannon is credited with developing theories on the storage of information and digital circuit design. His theories laid the foundation for the development of the Internet and modern computers.

ESTEEMED PERSONS OF MINORITY

✦ **Granville Tailer Woods** *(1856-1910)* is acknowledged as the first African-American electrical engineer. Having poor parents, Woods started his career working as an apprentice in a local machine shop in Ohio until he was able to attend college. He was also an inventor who held over 60 patents including the multiplex telegraph, telephone transmitter, and tunnel construction for the electric railroad system. The patent for the telephone transmitter was sold to the American Bell Telephone Company, which was later acquired by AT&T!

DID YOU KNOW?

The telegraph was the first practical application of electricity?

WOMEN IN THE FIELD

✦ **Edith Clarke** *(1883 – 1959)*, in 1918, became first woman to earn a degree in Electrical Engineering when she graduated from the Massachusetts Institute of Technology. She went on to invent the Clarke Calculator which is used to solve equations involving electrical current, voltage and impedance in power transmission lines. Being unable to find work as an engineer, she initially went on to work at General Electric and the University of Texas. Undaunted by these difficulties, she wrote *Circuit Analysis of A-C Power Systems,* a very influential textbook. Edith Clarke was posthumously inducted into the National Inventors Hall of Fame in 2005.

CURRENT PERCENTAGE OF WOMEN & MINORITY EMPLOYED IN THE FIELD

➡WOMEN 12%
➡WHITE 67%
➡ASIAN 17%
➡BLACK 5%
➡HISPANIC 9%

PROJECTED PERCENTAGE NEED FOR THE FUTURE

The employment, or size, of this occupation in 2020, which is the base year of the 2020-30 employment projections was 313,200. The projected numeric change in employment from 2020 to 2030 is 20,400 annually.

★ The projected percent change in employment from 2020 to 2030 is 7% (bls.gov)

ALTERNATIVE CURRENT (AC)

Nikola Tesla made this important electrical discovery in 1888. Alternative current (AC) differs from direct current (DC) in that the flow of electric charge periodically reverses over long-distance travel, making electricity accessible to all. AC is used for both domestic and commercial power.

RIO MADEIRA, BRAZIL

Transmission links are used to transport bulk electricity efficiently over a long distance. Brazilian energy providers Interligação hold the record of the world's longest transmission link, Elétrica do Madeira (IE Madeira). The 600kV Rio Madeira High Voltage Direct Current system spans 2375km which is about half the width of the United States. It brings electricity from hydropower plants on the Madeira River in the Amazon basin to the load centers in southeastern Brazil.

KINETIS K102

The Kinetis K102 is the world's smallest ARM-powered chip and a Microelectronics wonder. Manufactured by Freescale Semiconductors in the US, it measures just 2 x 2 x 0.5 millimeters (about as large as two ants side-by-side). This microchip is a full microcontroller unit with 4KB RAM, 32KB flash memory, a 32-bit 48 MHz processor, and a 12-bit analog-to-digital converter. It is a complete computer but so tiny that it can be swallowed! This microchip will revolutionize modern medicine.

ELECTRIC CARS

Electric cars are trending and becoming the preferred option for land travel. They are the most environmentally friendly mode of transportation, as the energy needed to run the engine is stored in batteries. It gives instant torque and smooth acceleration. The BMW i3, launched in 2014, has been certified by the Environmental Protection Agency (EPA) as the most fuel-efficient vehicle. The total fuel economy is calculated at 29 kW-hours per 100 miles, making it the first zero-emission mass-produced vehicle to use electric power.

THREE GORGES DAM

The Three Gorges Dam is a hydroelectric power plant classified as the world's largest dam. It spans 2,335 meters (about twice the height of the Burj Khalifa, the tallest building in the world), and is capable of producing 22,500 MW of energy. Located in China and spanning the Yangtze River, the dam was intended to protect millions of people from flooding that is common in the region. Construction completion was in 2006 but the full generating capacity of the turbines occurred in 2012.

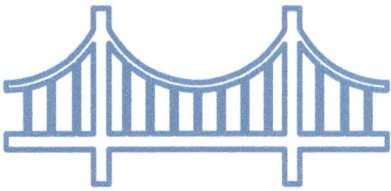

CIVIL ENGINEERING

DESCRIPTION OF FIELD

This field of Engineering is one of the two oldest specialties, having existed since the first civilian constructed a place to live, or created a means to get across a body of water. (The second oldest specialty is military engineering.) Both existed before the Industrial Revolution took place in the 18th century. It is said that all other specialties split from civil engineering.

The American Society of Civil Engineers (ASCE) was formed in 1852 and is the nation's oldest national engineering society. The field of civil engineering is concerned with the planning, designing, constructing, and maintaining of fixed or static structures such as bridges and buildings. Civil engineers are concerned with developing new ways to economically use forces and materials of nature to provide facilities for transportation, industry, and human satisfaction.

It was ranked third by the Bureau of Labor Statistics (BLS) in 2020 on the list of most in-demand engineering jobs.

The water slide was
designed by a
Civil Engineer?

WHAT DO THEY DO?

Civil Engineers manipulate geography such as earth and water to solve human problems.
Within civil engineering are other sub-specialties highlighted below:

➡Construction and Structural Engineering –
buildings; bridges; tunnels

➡Water References–irrigation; swamp draining;
controlling water bodies; water supply; sewage
disposal

47

- ➡ Environmental Engineering – preservation and cleanup
- ➡ Transportation Engineering – highway and railroad building and repair; traffic control
- ➡ Geotechnical Engineering – earthworks, that is, soil mechanics and foundations

WHERE DO THEY WORK?

Careers include employment by consulting and design companies, construction firms, local, state, and federal government agencies, research and development firms, and even law firms!

NOTABLE HISTORICAL PERSONS IN THIS FIELD

- ✦ **George John Smeaton** *(1729-1792)*, born in England, is considered to be the father of civil engineering and is even credited with coining the term. He worked on designs for bridges, canals, harbors and lighthouses. He was instrumental in the history, rediscovery and development of modern cement, and was the first person to use hydraulic lime in concrete.

- ✦ **Benjamin Wright** *(1770-1842)*, has been declared the father of American civil engineering. It was Wright who led the construction of the Erie, Delaware, Hudson, Chesapeake, and Ohio canals. Five of Wright's nine children followed in his footsteps to become civil engineers.

ESTEEMED PERSONS OF MINORITY

✦ **Howard P. Grant** *(1925-1997)* was the first black student to graduate from the University of California-Berkeley! He pursued a Bachelor of Science degree in Civil Engineering in 1948. Later that year, he became the first black member of the American Society of Civil Engineers (ASCE)

✦ **George Biddle Kelley** *(1884-1962)* graduated from Cornell University with a degree in Civil Engineering and became the first African American engineer to register with the state of New York. At Cornell, he co-founded and became the first President of the Alpha Phi Alpha Fraternity, Inc: the first African American fraternity. In the 1920's he was employed by the New York Engineering Dept where he worked on the NY State Canal system.

DID YOU KNOW?

Expansion joints are needed on roadways to allow expansion and contraction when the temperature fluctuates?

WOMEN IN THE FIELD

✦ **Hattie Peterson** *(1913-1993)* was the first Black female engineer to gain a Bachelor of Science in civil engineering. She became renowned when she became the first African American woman to join the U.S. Army Corps of Engineers (USACE) in 1954.

CURRENT PERCENTAGE OF WOMAN & MINORITY EMPLOYED IN THE FIELD

➡ 83.4% MALE

➡ 16.4% FEMALE

➡ 71% WHITE

➡ 11% ASIAN

➡ 5% BLACK

AVERAGE MALE SALARY

➡ $98,586

AVERAGE FEMALE SALARY

➡ $80,829

PROJECTED PERCENTAGE NEED FOR THE FUTURE.

The projected percent change in employment from 2020 to 2030 is 8%. The employment, or size, of this occupation in 2020, which is the base year of the 2020-30 employment projections was 309,800.

★ The projected numeric change in employment from 2020 to 2030 is 25,300 annually.

◄ EXAMPLES OF CIVIL ENGINEERING

THE BAILONG ELEVATOR (ZHANGJIAJIE, CHINA)

Built off the side of an enormous cliff in Zhangjiajie National Forest Park in China, the Bailong Elevator is the highest and heaviest outdoor elevator in the world. Enforced with three double deck glass elevators, it ascends to a maximum height of 326 meters (about the height of the Empire State Building) in a mere two minutes. It was completed in 2002.

CHANNEL TUNNEL

This is the world's longest underwater tunnel, that runs 32 miles connecting England and France beneath the English Channel. When it opened in 1994, it was the most expensive project of all time, with the final cost of an astounding $10 billion. The lowest part of the tunnel is 250 feet deep while 23.5 miles of the rail is submerged underwater. At speeds of 200 miles per hour it takes about 20 minutes to traverse the distance.

GREAT WALL OF CHINA

The Great Wall was originally constructed of compressed earth and was to protect China from Mongolian tribes. The Ming Dynasty fortified the wall between 1368 and 1644 by using granite and bricks, with certain strategic areas redesigned. The original wall runs for 5,500 miles (about twice the distance of California to Maine). Its construction happened over a 2000-year span.

DID YOU KNOW?

Concrete continues to dry and strengthen over it's lifespan?

HOOVER DAM

In one of the most significant undertakings of civil engineering, over 21,000 men (about the seating capacity of Madison Square Garden), worked on the Hoover Dam during The Great Depression, with an entire city – known as Boulder City – erected to house the workers. Today, Hoover Dam is considered an incredible legacy. It plays a vital role in flood prevention of the Colorado River, and its 17 turbines generate electricity that powers 1.3 million homes. Its role in American history, and its part in industrial evolution, make it a true wonder of the modern world.

ENVIRONMENTAL ENGINEERING

DESCRIPTION OF FIELD

This engineering specialty is focused on improving environmental quality and protecting individuals from adverse environmental issues like pollution. The field was born in the 19th century, out of people's need for clean water and sewage disposal as settlements grew larger.

It then grew with the establishment of manufacturing and agricultural plants that threatened to contaminate the soil and air. Environmental engineers apply their knowledge of soil science, water, biology, chemistry, and engineering to design, implement and maintain solutions to environmental problems.

This field was ranked by the Bureau of Labor Statistics (BLS) as the fourth most in-demand engineering job in 2020. The American Academy of Environmental Engineers (AAEE) was officially formed in 1973. The name was changed to American Academy of Environmental Engineers and Scientists (AAEES) in 2012.

DID YOU KNOW?

Groundwater contributes to about 30% of the drinking water in the United States?

WHAT DO THEY DO?

Environmental Engineers work to improve waste disposal, pollution (air & water), waste disposal, and recycling. They work towards improving public health, increasing recycling, improving waste disposal, and controlling air and water pollution, as per the U.S. Bureau of Labor Statistics. The activities of environmental engineers include, but are not limited to, the construction, planning, operation, and design of wastewater treatment facilities in industries and

municipalities. They also analyze the quality of groundwater and surface water; plan for the reuse and disposal of sludges and wastewater; collect, process, transport; and recover and dispose of solid wastes as per accepted engineering practices. They partner with businesses to identify references of contamination and pollution including ways to decrease or dispose of it safely.

WHERE DO THEY WORK?

Careers include working with government agencies to solve problems of clean water, recycling, and other departments. Their job can be office-based if they are designing and planning, or outdoors if they are carrying out solutions and doing assessments.

NOTABLE HISTORICAL PERSONS IN THIS FIELD

+ **Joseph Bazalgette** *(1819-1891)* supervised the construction of the first municipal sanitary sewer system in London in the 19th century. In that time, there was an outbreak of cholera due to raw sewage being dumped into the Thames River, which was the primary source of drinking water.

+ **Marc Edwards** *(born in 1964)*, is a civil and environmental engineer. He is a Professor at Virginia Tech and an expert on corrosion and water treatment. He did a research project, which led to the detection and reduction of lead in drinking water in the US.

ESTEEMED PERSONS OF MINORITY

✦ **Lilia A. Abron** *(born in 1945)* is a chemical engineer and entrepreneur. Inspired by Rachel Carson and the growing environmental movement, Abron also pursued a career in environmental engineering. Abron's doctoral thesis was on the removal of pesticides from water. In 1978 she incorporated PEERCP, an engineering consulting firm focusing on environmentally sustainable solutions. This made her the first African American to start an environmental engineering company.

DID YOU KNOW ?

The Romans used the force of gravity to move water as far as 57 miles?

WOMEN IN THE FIELD

✦ **Ellen Henrietta Swallow Richards** *(1842-1911):* Her work lay necessary foundations and contributed advances in plumbing, industrial health and safety. She was also the first woman to be accepted in, and receive a degree from a technical university in America. In 1887 she led the first comprehensive water-quality study in the US, which led to the nation's first water-quality standards and wastewater treatment facility.

✦ **Krystle McClain** has a master's degree in environmental engineering and won the 2019 Black Engineers Award in Professional Achievement. She is an African American woman, employed at Naval Facilities Engineering Command Far East, where she is Commander of Fleet Activities in Yokosuka, Japan. This base is the largest US overseas naval base.

CURRENT PERCENTAGE OF WOMAN & MINORITY EMPLOYED IN THE FIELD

➡ 70.2% MALE

➡ 28.7% FEMALE (SOURCE: DATA USA)

➡ 79% WHITE

➡ 12% ASIAN

➡ 3% BLACK

AVERAGE MALE SALARY

➡ $97,857

AVERAGE FEMALE SALARY

➡ $80,091

PROJECTED PERCENTAGE NEED FOR THE FUTURE.

The projected percent change in employment from 2020 to 2030 is 4% (bls.gov)

The employment, or size, of this occupation in 2019 was 30,200 (Census Bureau)

★ The projected numeric change in employment from 2020 to 2030 is 1,900 annually.

◄ EXAMPLES OF ENVIRONMENTAL ENGINEERING

EASTGATE BUILDING (HARARE, ZIMBABWE)

The Eastgate Building is widely thought to be the first building of its caliber to utilize a natural ventilation system incorporating the heat-regulating technology found in the towering termite mounds of Southern Africa. The conical termite mounds, which can grow to several meters in height, maintain a nearly constant internal temperature while exterior conditions swing from 108F to 37F (42C to 3C).

Similar to the mounds, the materials that constructed the Eastgate Building have a high "thermal mass" enabling the structure to absorb high heat without changing temperature. Architect Mick Pearce and engineers from Arup dreamed up the design to use a system of fans, vents, and funnels. The office complex, therefore, uses only 10% as much energy as similarly sized buildings.

DID YOU KNOW ?

Clouds are made up of water droplets that are much smaller than a raindrop?

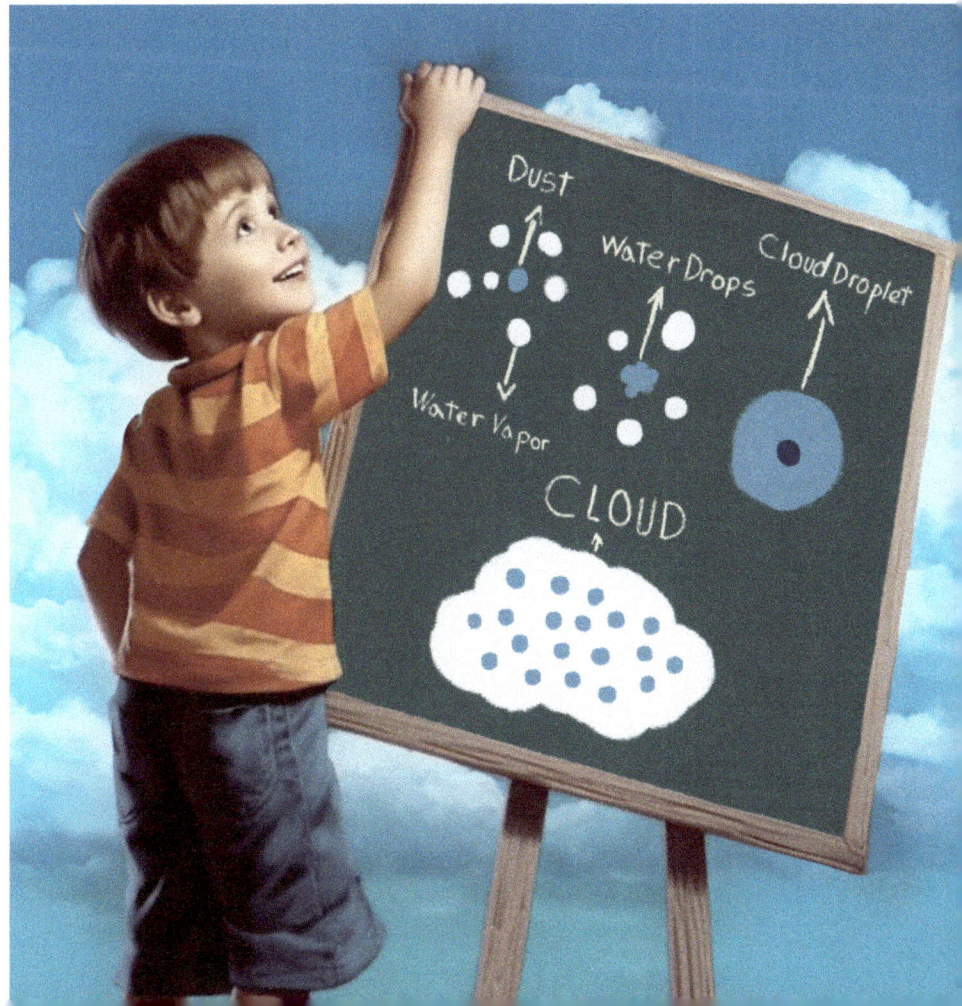

MILLENNIUM DOME (LONDON, ENGLAND)

Millennium Dome (later re-named The O2), is a concert and sports venue that is 52 meters at its highest point. It encompasses a sprawling and virtually uninterrupted internal space using remarkably little material: 1-2 pounds per square meter compared to the 30-40 pounds typical of most roofs. The building has twelve steel masts (one for each month), each towering to 100 meters (about the length of a football field), and together supporting a Teflon-coated, glass-fiber roof of over 100,000 square meters.

JAPAN PAVILION FOR EXPO 2000 (HANNOVER, GERMANY)

Representing sustainable architecture, using recycled materials was the theme of the Japan Pavilion for Expo 2000. It was designed to reduce the environmental impact and reduce industrial waste when taken apart. This structure had an undulating tunnel arch — a grid of gently swooping paper tubes covered by a paper membrane and supported by pulling cables — (measuring 73.8 x 25 x 15.9 meters) and featured a wooden arch for strength at each end.

The Aqueduct was built by the Romans in as early as 312 BCE?

CHARLES DAVID KEELING APARTMENTS (SAN DIEGO, CALIFORNIA)

The Charles David Keeling Apartments are designed to capture the nearby ocean breeze thereby promoting natural ventilation and reducing energy consumption. The buildings also include solar cells and a number of features including on-site wastewater systems and other low-impact development features.

AUTONOMOUS POWERBUOY

The Autonomous PowerBuoy from Ocean Power Technologies features power harvesting and tracking capabilities, as well as vessel detection of surface and subsurface vessels. This PowerBuoy utilizes clean energy and acts as an uninterruptable power supply. It recharges itself by harvesting wave energy and may be used for a variety of purposes including security, fish farming as well as oil and gas operations.

MATERIALS ENGINEERING

DESCRIPTION OF FIELD

Filed of Engineers that apply their knowledge of ceramics, composites, metals, plastics, composites, nanomaterials and other substances to create new materials. They may use computers to design and model their products. Their work greatly impacts other fields of engineering, as their understanding of materials supports other specialties. The Amcrican Society for Materials (ASM) International was formed in 1913.

WHAT DO THEY DO?

Materials Engineers develop, process, and test materials that are used to create a wide variety of products such as memory chips to sporting equipment, vehicle parts and biomedical devices.

WHERE DO THEY WORK?

Careers include employment in transportation equipment manufacturing, factories, and research and development facilities. They also work with other engineers in different specialties.

NOTABLE HISTORICAL PERSONS IN THIS FIELD

✦ **Morris E. Fine** *(1918-2015)* was Professor Emeritus of Materials Science and Engineering at Northwestern University in Evanston, Illinois. He is credited as being co-founder of the world's first materials science department at the institution. He is widely regarded as a father of materials science, being knowledgeable about metals, alloys, ceramics, and composite materials. His research, books, and practical contributions are acknowledged worldwide.

ESTEEMED PERSONS OF MINORITY

✦ **Clinique L. Brundidge** *(born in 1984)* became the first Black woman to receive a Ph.D. in Materials Science and Engineering from the University of Michigan. Her father worked as a manufacturing engineer at General Motors. She would visit him at the plants and over time grew curious about metallurgy and then principles of mechanical engineering. She now works in the design and development of nuclear core reactors, for nuclear-powered war ships. Currently, she is working as a Principal Engineer at the Naval Nuclear Laboratory in Pittsburgh, Pennsylvania.

WOMEN IN THE FIELD

✦ **Stephanie Kwolek** *(1923-2014)* is credited with the discovery of Kevlar, a synthetic material five times as strong as steel. It is used in the construction of bullet-proof vests, helmets, cables and even camping gear!

DID YOU KNOW ?

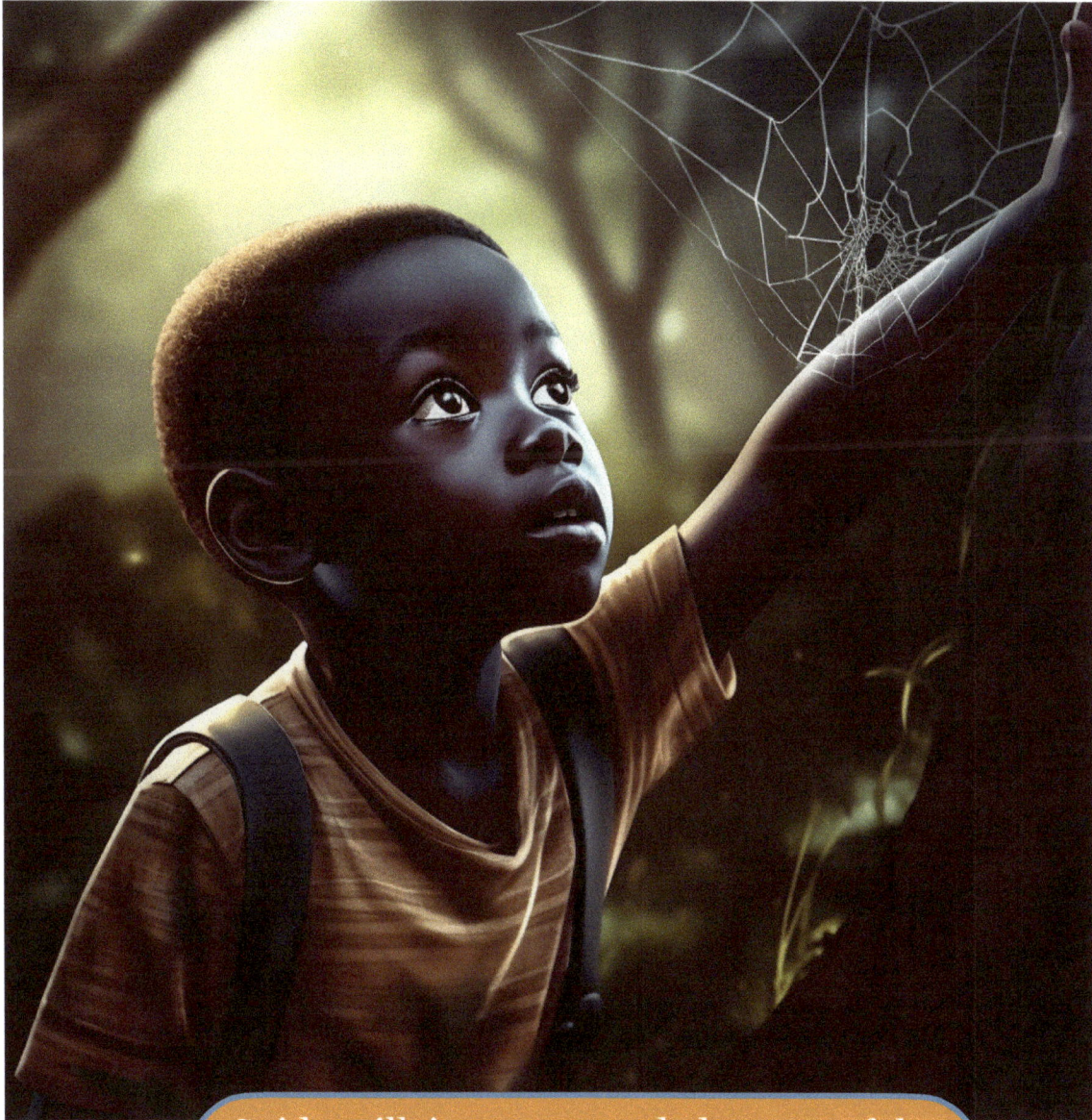

Spider silk is now regarded as one of the strongest materials in the world?

CURRENT PERCENTAGE OF WOMAN & MINORITY EMPLOYED IN THE FIELD

- ➡ 84.2% MALE
- ➡ 15.7% FEMALE
- ➡ 75% WHITE
- ➡ 12% ASIAN
- ➡ 5% BLACK

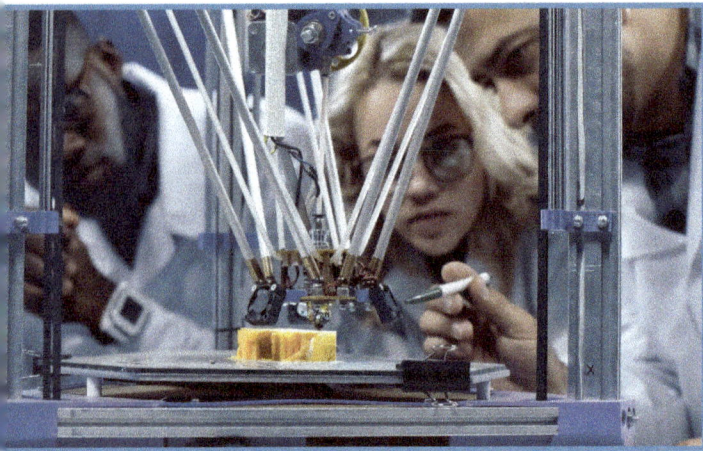

AVERAGE MALE SALARY

➡ $87,978

AVERAGE FEMALE SALARY

➡$91,971

PROJECTED PERCENTAGE NEED FOR THE FUTURE

The projected percent change in employment from 2020 to 2030 is 8% . The employment, or size, of this occupation in 2020 was 25,100

★ The projected numeric change in employment from 2020 to 2030 is 2,100 annually.

EXAMPLES OF MATERIAL ENGINEERING

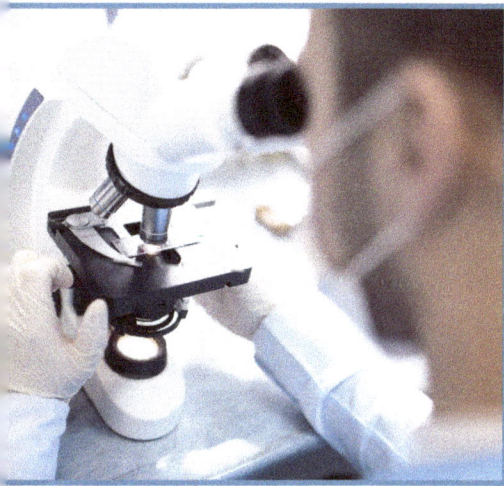

SCANNING PROBE MICROSCOPES

The invention of the scanning tunneling microscope (STM) credited to IBM scientists, opened doors into the world of nanotechnology and other diverse sectors. Using this device scientists were able to go beyond the surface of solids because a look into the atomic structure of particles was now possible. An Electron microscope was born — winning the Nobel Peace Prize in 1986.

DID YOU KNOW?

Gold platinum alloys are engineered materials that prove to be 100 times more abrasion resistant than high-strength steel?

GIANT MAGNETORESISTIVE EFFECT

In 1988 a phenomenon was observed by French and German scientists between magnetic. They discovered the large change in electrical resistance in some materials in reaction to the magnetic field employed. This is the main technology behind how hard discs and biosensors are read.

SEMICONDUCTOR LASERS AND LEDS

Semiconductor lasers and LEDs are light references. They emit diodes used for illumination and vary in wavelengths. These creations paved the way for fiber optic communication, along with other electronic devices – CDs, laser printing and the like.

MATERIALS FOR LITHIUM ION BATTERIES

These are rechargeable batteries that can store elevated levels of power. The lithium-ion battery is used in smartphones and many other devices. They can be charged and discharged again and again. The lithium ions move between negative and positive electrodes called conductors.

CARBON NANOTUBES

Carbon Nanotubes are hollow tubes made up of carbon atoms (graphene). They may be single-walled or multi-walled. Carbon nanotubes are classified as the strongest and stiffest of all materials. Their discovery was attributed to Sumio Iijima of NEC, Japan in 1991.

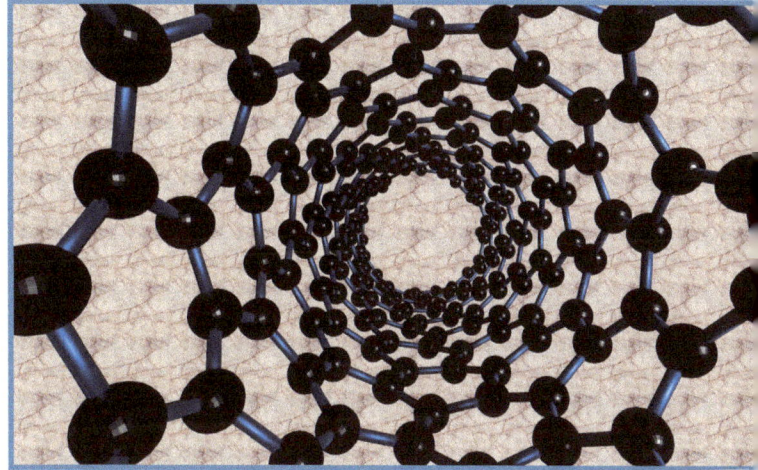

This material is widely used for its electrical conductivity, high strength, and thermal conductivity. Today this material is present in sporting goods, bulletproof vests, and space materials.

BIOMEDICAL ENGINEERING

DESCRIPTION OF FIELD

This specialty of engineering consists of both engineering science and applied engineering for the purpose of defining and solving problems in the field of medicine. This directly impacts medical research and clinical medicine and thus improves healthcare in general. In addition to the typical areas of mathematics and physics, biomedical engineers must be competent in anatomy, medicine, and physiology.

Most persons employed in this field must have sound knowledge in another engineering specialty such as mechanical, chemical, or electrical so they may apply these principles to understanding and manipulating biological systems. Biomedical engineering was ranked as the fifth most in-demand engineering job in 2020.

The field traces back to the 17th century, but the Biomedical Engineering Society (BMES) was not formed until 1968.

DID YOU KNOW ?

Studying the digestive tract of a mouse can help explain the break down biomass for use in bioproducts?

WHAT DO THEY DO?

Biomedical Engineers combine their knowledge of biology and engineering to develop devices and procedures. Options in this field include:

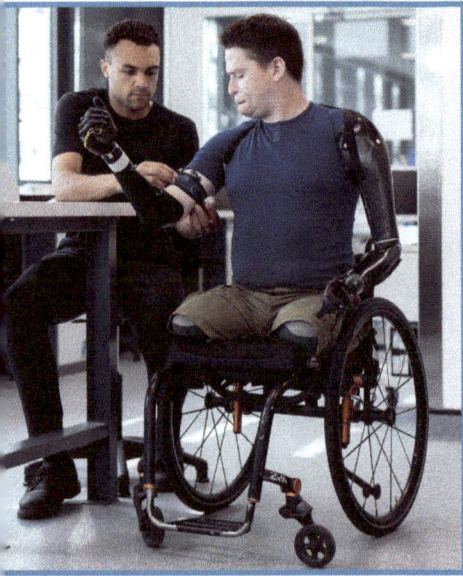

➡Doing research to develop products such as artificial joints, prostheses and medical information systems.

➡Design external medical devices such as surgical robotics tools, magnetic resonance imaging (MRI), dialysis machines

➡Design, develop and maintain implanted devices such as pacemakers, dentures and artificial organs

➡Design safety devices such the bullet-proof vest for the security forces and fire suits for firefighters

Sub-specialties include biomaterials, orthopedic engineering, biomechanics and rehabilitation engineering.

WHERE DO THEY WORK?

Biomedical engineers are employed at universities, research facilities, medical institutions, government agencies, and pharmaceutical companies. Some reputable employers are Johnson & Johnson, Mayo Clinic, Blatchford, and Genzyme.

DID YOU KNOW?

National Institute of Biomedical Imaging and Bioengineering is the primary government agency in the United States for biomedical and public health research?

NOTABLE HISTORICAL PERSONS IN THIS FIELD

✦ **Robert Koffler Jarvik** *(born in 1946)* is an American scientist known for inventing the first permanently implantable artificial heart. As a teenager he developed the surgical stapler and multiple other tools used in medicine.

ESTEEMED PERSONS OF MINORITY

✦ **Raphael Carl Lee** *(born in 1949)* is the Paul and Allene Russell Professor of Surgery at the University of Chicago. In 2018, the American Institute of Medical and Biological Engineering awarded him the Pierre Galletti Award - the highest award in biomedical engineering in the U.S.

✦ **Treena Livingston Arinzeh** *(born in 1970)* boasts a master's in biomedical engineering from Johns Hopkins University. In 2003, she became the first individual to demonstrate that scientists can form viable, functional bone tissue which is not rejected by the body by implanting donor stem cells that are derived from the bone marrow of adults.

WOMEN IN THE FIELD

✦ **Dr. Nirmala 'Nimmi' Ramanujam** is Malaysian by birth, but migrated to the US for college. Dr. Ramanujam is a notable engineer and Distinguished Professor of Biomedical Engineering at Duke University (second highest ranked biomedical engineering program in the U.S.). She led a team to invent the pocket colposcope and callascope which are used in cervical cancer screening.

CURRENT PERCENTAGE OF WOMAN & MINORITY EMPLOYED IN THE FIELD

➡ 67.9% MALE

➡ 25.1% FEMALE

➡ 59.1% WHITE

➡ 19.8% ASIAN

➡ 12.7%; HISPANIC

➡ 4.7% BLACK

AVERAGE SALARY

$78,319 to $111,000

DID YOU KNOW?

The Abiocor Artificial Heart is completely self-contained inside the body?

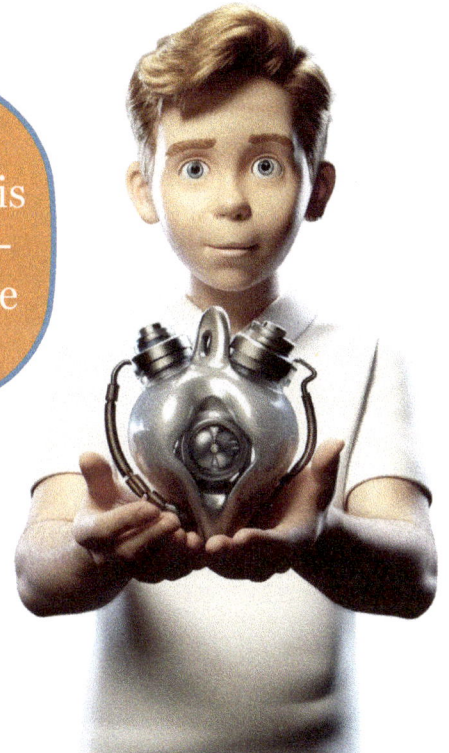

PROJECTED PERCENTAGE NEED FOR THE FUTURE.

The projected percent change in employment from 2020 to 2030 is 6% . The employment, or size, of this occupation in 2020 was 19,300.

★ The projected numeric change in employment from 2020 to 2030 is 1,100 annually.

EXAMPLES OF BIOMEDICAL ENGINEERING

THE ROBOT SURGEON

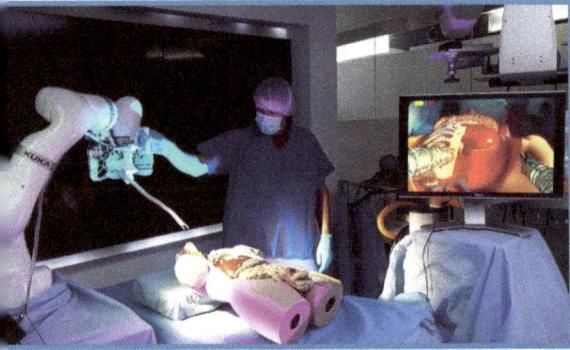

When it comes to diagnosing the patient, Artificial Intelligence (AI) is breaking new ground. Today, robotic surgery is on the rise in many hospitals. The robot surgeon can make smaller incisions and perform more precise and complex procedures. The benefits are a greater range of motion, less risk of infections and less blood loss.

GENOME EDITING (CRISPR-CAS9)

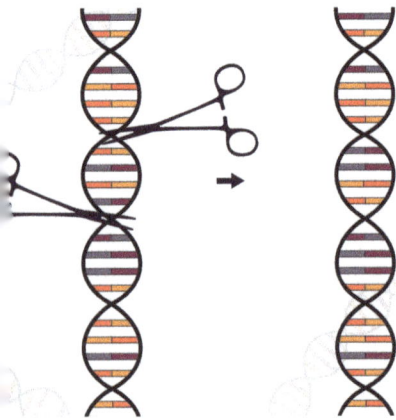

Genome Editing is no longer just a wish — technology has allowed this to be an option for anyone who wants to make changes to DNA. Genome editing enables you to control physical changes such as eye color or risk of disease. This is applicable to plants, animals and bacteria. CRISPR is an editing tool that makes this easy. A genome editing tool called TALENs was used on a human infant in 2015 to help fight cancer. Laws on genome editing are set by countries as there are ethical concerns.

ADVANCED PROSTHETICS

Prosthetics have become a necessity for some, and that means that they need to work really well. Bionic limbs use signals from the person's muscles to be able to integrate with muscles and mimic human limbs. The integration happens by reading muscles' contraction and nerve electrodes to improve sensation and command. This is no longer just a medical process but a combination of medical and technological advancements.

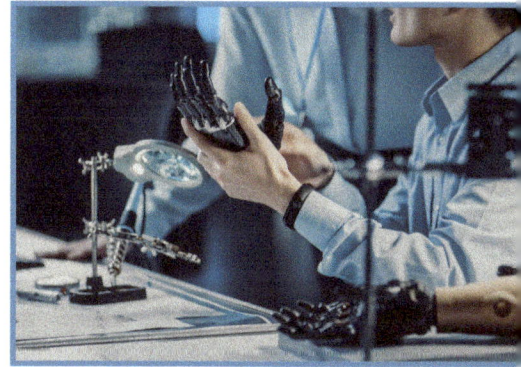

DID YOU KNOW?

With the aid of the 3D printer, bio engineers are able to grow viable organs for humans?

BIONIC CONTACT LENS

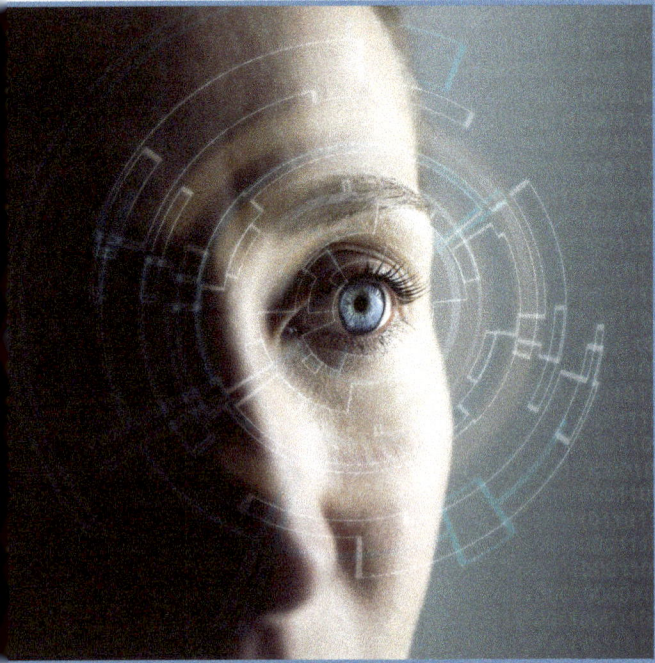

The bionic contact lens is in development and will be instrumental in changing the way information is gathered for the treatment and diagnosis of illnesses. The lens work by overlaying images, information, and notifications onto the real world. The smart contact lens could, for instance, send its wearer low blood glucose alerts if they are diabetic, or enable them to take pictures or record videos just by blinking.

AGRICULTURAL ENGINEERING

DESCRIPTION OF FIELD

This field of engineering is concerned with land development, soil management, the design of farm machinery, as well as processes related to livestock and crop farming. Agricultural engineers use their engineering skills to solve agricultural problems and improve agricultural processes. The American Society of Agricultural Engineers (ASAE) was formed in 1907. The society's name changed to the American Society of Agricultural and Biological Engineers (ASABE) in 2005.

WHAT DO THEY DO?

Agricultural engineers influence the storage and processing of agricultural crops as well as tending to pollution and environmental issues and the use of structures and facilities.

Their work may be in farming land products, aquaculture, forestry, and food processing. Roles that are related to these such as tending to livestock, disposing of animal waste, fertilizer application, and harvesting systems would also be influenced by agricultural engineers.

DID YOU KNOW ?

Agricultural engineers solve farming problems?

WHERE DO THEY WORK?

Careers include employment in the following industries: federal government; consulting firms; other engineering firms; research facilities and machinery manufacturing.

NOTABLE HISTORICAL PERSONS IN THIS FIELD

✦ **Jay Brownlee Davidson** *(1880-1957)* was a university professor, agricultural engineer, and consultant. He was the founder and first president of the American Society of Agricultural and Biological Engineers (then known as the American Society of Agricultural Engineers, ASAE), and is considered to be the father of agricultural engineering.

WOMEN IN THE FIELD

✦ **Monica Lozano Luque** earned a degree in agricultural engineering from EARTH University in Costa Rica. She led a national organic farming program in Colombia and, in 2006, incorporated Sea Soil, S.A. in Bogota, Colombia. Sea Soil assists organizations in transitioning to strategies that are more environmentally sustainable in nature. Also, it imports technology for the improvement of the fertility of plants and soil. Being the consulting partnership's technical director, Luque provides tailored services to producers to improve natural resource management.

✦ **Susana Chaves Villalobos** is a Costa Rican agricultural engineer. She is the founder of IBS Soluciones Verdes, which assists small scale farmers with production certification, marketing, strategies, and communication. Additionally, she runs the Yo Como Verde (I Eat Green) campaign, thereby helping in the promotion of healthy eating in Costa Rica. Furthermore, she has the accreditation to train and certify organic farm production.

DID YOU KNOW?

That farmers produce more than enough food to feed the world's population of 7.6 billion people? In fact, in total, farmers produce 1.5 times what is needed!

CURRENT PERCENTAGE OF WOMAN & MINORITY EMPLOYED IN THE FIELD

- ➡ 73% MALE
- ➡ 20% FEMALE
- ➡ 62.8% WHITE
- ➡ 13.5% ASIAN
- ➡ 14.9% HISPANIC
- ➡ 5.1% BLACK

MALE AVERAGE SALARY

- ➡ $65,734

FEMALE AVERAGE SALARY

- ➡ $65,143

PROJECTED PERCENTAGE NEED FOR THE FUTURE

The projected percent change in employment from 2020 to 2030 is 5%. The employment, or size, of this occupation in 2020 was 1,500.

★ The projected numeric change in employment from 2020 to 2030 is 100 annually.

EXAMPLES OF AGRICULTURAL ENGINEERING

BEES & DRONES

The extinction of bees will have a devastating effect on agriculture and the world's economy. Coupled with the realities of climate change, scientists are concerned that food shortages will affect fair prices. Due to the decline of the bee population, drones have taken up their primary duty of aiding in the pollination process. Multiple versions of these Drones now exist on different scales.

VERTICAL FARMING

Vertical Gardening, Hydroponics and Aeroponics are leading the way for future farming. As the task of feeding the world is getting more complex with real threats of global warming, different techniques will need to be employed for the successful facilitation of this task. Both *hydroponics* and *aeroponics* have less environmental impact than traditional farming. Coupled with a vertical growing space, the scale of farming yield is made unlimited even with urban shrinkage.

TUNNEL BORER AUTOMATION

Tunnel boring started in 1825 with the building of the Thames. It is still expensive and dangerous. The automation of this process increases precision, reliability, and efficiency. Engineering has recently added artificial intelligence components to tunnel boring machines, allowing these machines to be autonomous with the ability to self-correct during the boring process.

STRUCTURAL ENGINEERING

DESCRIPTION OF FIELD

This field of engineering is a sub-specialty of civil engineering. Structural engineers are concerned with the design and planning of buildings and other structures, to ensure they are safe, strong, stable, and able to manage loads. Structural engineering also takes in checking the earthquake-susceptibility, of structures. They work closely with architects.

The Structural Engineering Institute (SEI), formed in 1996, is a sub-group within the American Society of Civil Engineers (ASCE).

WHAT DO THEY DO?

Structural engineers are trained professionals who apply their technical knowledge to specify various kinds of construction materials in different geometries and design structures and shapes which can withstand environmental stresses and pressures like storms, earthquakes and gravity loads. Structural Engineering includes:

➡Earthquake engineering.
➡Façade engineering.
➡Fire engineering.
➡Roof engineering.
➡Tower engineering.
➡Wind engineering.

87

DID YOU KNOW ?

Structural engineering is one of the oldest types of engineering?

WHERE DO THEY WORK?

Career options for structural engineers include work in construction and engineering firms. These individuals are integrated into a project if an owner is planning to change the building's usage. This may include introducing additional floors or considerably expanding the building. Also, these engineers are contacted if there is any damage to a structure due to environmental deterioration, fire, corrosion, wear and tear, or impact. Any of these structural issues could lead to capacity loss and pose a threat to safety.

NOTABLE HISTORICAL PERSONS IN THIS FIELD

✦ **Fazlur Rahman Khan** *(1929-1982)* was a Bangladeshi-American structural engineer and architect who pioneered computer-aided design (CAD) and important structural systems for skyscrapers. He is regarded as the father of tubular designs for skyscrapers and is considered by many to be the greatest structural engineer of the 20th century. He contributed to multiple skyscraper projects including Chicago's John Hancock Center and the Sears (now Willis) Tower.

✦ **Ove Arup** *(1895-1988)* was the Anlo-Danish engineer who founded Arup Group Limited, which is one of the foremost engineering firms in the world. He is widely considered the foremost engineer of his time. His most legendary works include the Van Ginkel Footbridge and the Sydney Opera House.

ESTEEMED PERSONS OF MINORITY

Charles Sumner-Duke *(1879-1952)* was an architect and structural engineer. Duke was also a public official who sought to help African Americans access opportunities. He co-founded the National Technical Association (NTA) in 1925.

WOMEN IN THE FIELD

✦ **Ruth Gordon Schnapp** *(1926-2014)* is known widely as the first female structural engineer. She graduated from Stanford University in 1950 with a master's degree in structural engineering. In 1959, Schnapp later became the first female to be a state-certified structural engineer. She founded Pegasus Engineering, Inc. in 1984 and conducted safety earthquake survivability studies as well as post-earthquake evaluations in hospitals and schools.

✦ **Aine Brazil** is the vice-chairman of the engineering firm Thornton Tomasetti, and has overseen remarkable ways that have opened doors to the construction of some of the tallest and most unique infrastructure projects in the world. Aine was the lead structural engineer for 11 Times Square in New York, which is a 38-story, 975,000-square-foot tower. Currently, she is working on the Hudson Yards development that will float six city blocks over a train yard by using a concrete 'apron'.

CURRENT PERCENTAGE OF WOMAN & MINORITY EMPLOYED IN THE FIELD

➡ 13.7% WOMEN
➡ 68.2% WHITE
➡ 9.2% LATINO or HISPANICS
➡ 16.1% ASIANS

EXAMPLES OF STRUCTURAL ENGINEERING

COLOSSEUM (ROME, ITALY)

The Colosseum structure was finally completed and dedicated in 80 CE. This amphitheater could host 10,000 Romans. This historic site is a visual representation of ancient Rome's architectural and engineering fortitude. Originally having four stories, the structure stands tall at a height of 159 feet. The present dilapidated condition of the monument occurred as a result of earthquakes, in addition to being abandoned after being used for four generations. Preservation of the Colosseum began in the 19th century.

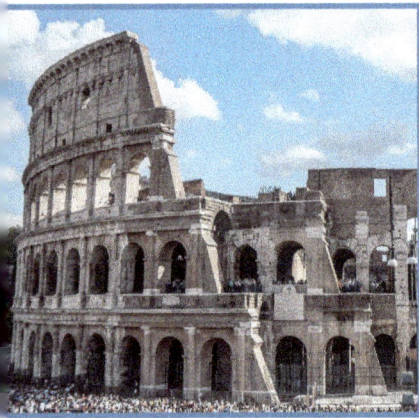

STONEHENGE (SALISBURY, ENGLAND)

Stonehenge — a ring of stones constructed 4000-5000 years ago, is a circular archeological site, estimated to be built in about 1520 CE at the end of the Stone Age. The uprights measure 5.5 meters high and weigh as much as 45 tons. Today Stonehenge lends homage to the mystery and power of our prehistoric past.

SYDNEY OPERA HOUSE (AUSTRALIA)

Designed by Jørn Utzon and engineered by the founder of the world-renowned company Arup, the Sydney Opera House in Australia finally opened in 1973. The project completion was delayed due to construction and engineering challenges of the precast concrete shells which took 4 years to solve. The structure. Sir Ove, the founder of Arup is quoted as saying, *"Engineering problems are under-defined, there are many solutions, good, bad and indifferent. The art is to arrive at a good solution. This is a creative activity, involving imagination, intuition, and deliberate choice."*

HAOHAN QIAO BRIDGE (CHINA)

Located in the Shiniuzhai National Geological Park, the Brave Man's bridge is a 984 feet long and 590 feet high, all-glass suspension bridge. So that means that while you set foot on the bridge high in the air, you will clearly see how high you are walking. Interesting, isn't it? Absolutely not! It is greatly intimidating. Engineers replaced the wooden slats with reinforced glass (25 times stronger than normal glass) in 2014.

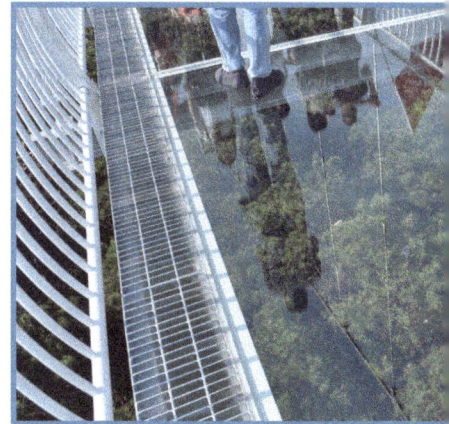

BURJ KHALIFA (DUBAI)

Burj Khalifa is currently hailed as the tallest building in the world. It was opened in 2014 and measures 828 meters high. This building was designed and engineered by a Chicago-based architectural firm, and sports 162 stories. Built with reinforced concrete and surrounded by glass, the Burj Khalifa is unprecedented, using geometric design and technology to withstand wind load and other dessert conditions.

ARCHITECTURAL ENGINEERING

DESCRIPTION OF FIELD

Architectural engineers apply the theory and practice of engineering and technological skills to design, test, and construct buildings. While architects are specialists in the stylistic, aesthetic, and functional aspects of design, architectural engineers apply principles of engineering to design and construct systems within a building. Architectural engineers integrate electrical, mechanical, lighting, acoustics, structural, and fire protection concepts into their work.

WHAT DO THEY DO?

Architectural engineers apply theoretical and practical knowledge to the design of building systems and buildings. The main aim is to engineer high-performance buildings which are economically viable, sustainable, resilient and ensure the comfort, productivity, safety and health of occupants.

WHERE DO THEY WORK?

There are broad-ranging employment opportunities for architectural engineers. These include consulting engineering firms, architectural engineering firms, producers and designers of building materials, real estate developers, building equipment designers, building owners, engineering and management groups, building technology consultants, specialty contractors, construction managers, contractors, and research facilities.

DID YOU KNOW?

The Empire State building generates a significant amount of money from its observational decks?

NOTABLE HISTORICAL PERSONS IN THIS FIELD

✦**Alexandre Gustave Eiffel** *(1832 - 1923)* was the man behind the church of Notre Dame Des Champs, Statue of Liberty, Eiffel Tower, and several other remarkable structures. He had an excellent understanding of science and mathematics, and on the Eiffel Tower, he calculated the distance between the riveted holes to within one-tenth of a millimeter.

✦ **Richard Buckminster Fuller** *(1895-1983)* was well-known for his invention of the geodesic dome, a semi-spherical structure that uses immensely sturdy self-bracing triangles. These structures are practically indestructible and lightweight (a segment measuring 15 feet and weighing around 4 pounds).

ESTEEMED PERSONS OF MINORITY

✦ **Moses McKissack** *(1879-1952)* cofounded *McKissack & McKissack* — the first Black-owned architectural firm in the United States. It is the oldest Black-owned architecture and engineering firm in the country. He became the first licensed architect in the southern states. McKissack's firm was contracted to build the 99th Pursuit Squadron Airbase in Tuskegee, Alabama. This was the largest federal contract at that time to be given to a Black-owned firm.

WOMEN IN THE FIELD

✦ **Beverly Loraine Greene** *(1915–1957)* was the first Black woman to be licensed as an architect in the United States. She graduated from the University of Illinois at Urbana-Champaign. Here, she became the first African-American woman to receive an architectural engineering degree. Following her graduation, she received a master's in City Planning and Housing.

CURRENT PERCENTAGE OF WOMAN & MINORITY EMPLOYED IN THE FIELD

➡ 83.5% MALE

➡ 16.7% FEMALE

➡ 69.7% WHITE

➡ 12.8% ASIAN

➡ 2%; HISPANIC

➡ 5.3% BLACK

MALE AVERAGE SALARY

➡ $93,753

FEMALE AVERAGE SALARY

➡ $77,323

PROJECTED PERCENTAGE NEED FOR THE FUTURE

Employment for architectural engineers will likely grow at a 3% rate from 2018 to 2028.

EXAMPLES OF ARCHITECTURAL ENGINEERING

MILLAU VIADUCT (FRANCE)

The Millau Viaduct holds the record as the world's tallest bridge with a height of 343 meters and a length of 2460 meters. It was completed in an astonishingly short time of 3 years being ahead of schedule. Completed in 2004, this cable-stayed road bridge connects Paris to Barcelona spanning the length of 30 city blocks. The project cost was nearly $450 Million, built with a life span of 120 years.

DID YOU KNOW?

Over 2.3 million blocks of stone were used to build the Great Pyramid of Khufu in Giza, Egypt?

THE VENICE TIDE BARRIER

The Venice Tide Barrier is the world's largest flood prevention project and was the solution to Venice's flooding prevention problem. For 40 years there was debate and discourse on how to prevent the city of Venice from sinking and keep it safe from flooding. The project started in 2003 and consists of 78 rotating gates, each with an area of 6,500 square feet (about twice the area of a tennis court).

The gates are large metal boxes that rest at the bottom of the sea. Whenever the tide rises above 3 1/2 feet, the water is removed from the gates, causing them to float.

BEIJING NATIONAL STADIUM (CHINA)

Built and used to stage the 2008 Summer Olympics, this structure has 80,000 permanent seats. It was designed by Arup architects and engineers and boasts 26 miles of concrete and steel. This complex design may look casual, but elements of advanced geometry are at play. The stadium was designed and engineered by Arup to withstand an 8.0 magnitude earthquake due to having flexible joints in multiple sections.

PETRONAS TOWERS (KUALA LUMPUR, MALAYSIA)

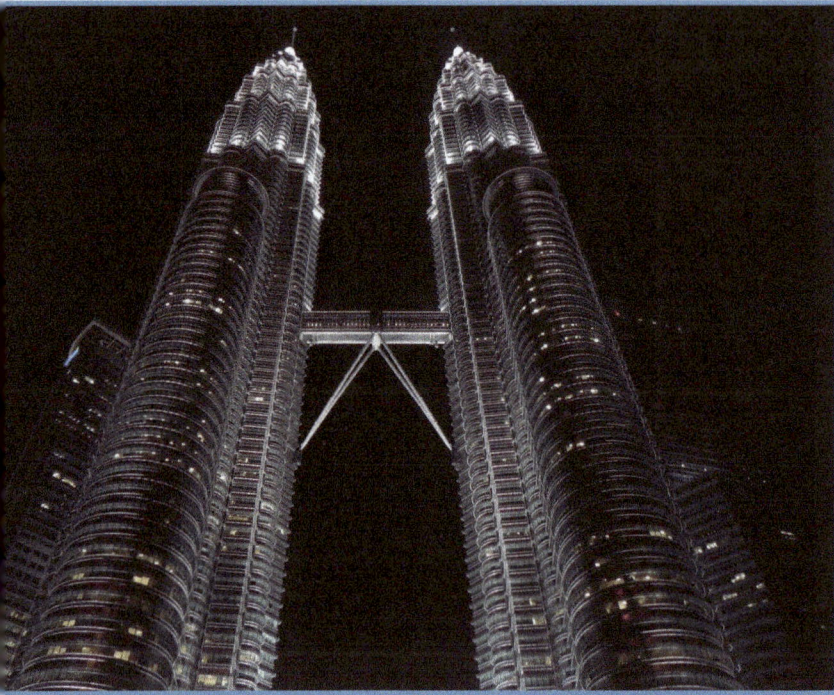

Standing tall at a height of 452m (about the height of the Empire State Building), the Petronas Towers is the tallest twin tower in the world. The towers also feature a 58.4-meters long bridge that connects the two towers. The glass panels of the twin towers have special properties such as light filtering and noise reduction. The twin towers are a representation of cultural and economic growth in Malaysia.

COMPUTER ENGINEERING

DESCRIPTION OF FIELD

Computer engineers are those who design, implement and test computer systems to solve problems, and improve computer systems' reliability and efficiency. The technology they design and develop is used widely such as in cars, aircraft, cell phones, or security systems.

DID YOU KNOW?

The basic function of a computer is to read, process, display and store data?

WHAT DO THEY DO?

Computer engineers work as software engineers, computer programmers, computer hardware engineers or computer systems analysts.

WHERE DO THEY WORK?

Careers include employment at research laboratories, digital consulting firms, technology manufacturers, financial institutions, telecommunication companies, government agencies, and security firms. They could also work in manufacturing, aeronautical, agricultural, or other industries instead of technology.

DID YOU KNOW?

Edgar Codd in 1949 invented the technique of multitasking, which allowed several computer programs to run at once?

NOTABLE HISTORICAL PERSONS IN THIS FIELD

✦ **Henry Edward Roberts** *(1941 – 2010)* was an American engineer, entrepreneur, and medical doctor, credited with inventing the first commercially successful personal computer in 1974.

ESTEEMED PERSONS OF MINORITY

✦ **Roy L. Clay Sr.** *(born in 1929)* was a computer programmer widely referred to as the Godfather of Silicon Valley. He spearheaded the development of HP's innovative computer, the 2116A. He also developed the software for the computer and led HP's Research and Development Computer group.

WOMEN IN THE FIELD

✦ **Masako Wakamiya** *(born in 1936)* is the world's oldest programmer and an iOS developer. She was a bank clerk before retirement. She bought her first computer and started to learn programming at 60 years old after she retired.

✦ **Augusta Ada Lovelace** *(1815-1852)* is believed to be the world's first computer programmer. She worked closely with Charles Babbage, the mathematician, in the 19th century. Lovelace wrote the world's first computer program.

CURRENT PERCENTAGE OF WOMAN & MINORITY EMPLOYED IN THE FIELD

➡ 73.2% MALE

➡ 27.4% FEMALE (SOURCE: datausa.io)

➡ 64.2% WHITE

➡ 17.2% ASIAN

➡ 1.7% HISPANIC

➡ 6.8% BLACK

MALE AVERAGE SALARY

➡ $94,923

FEMALE AVERAGE SALARY

➡ $75,476

DID YOU KNOW ?

Artificial intelligence became formal in 1956?
AI simply means "intelligence in machines."

PROJECTED PERCENTAGE NEED FOR THE FUTURE

The projected percent change in employment for computer hardware engineers from 2020 to 2030 is 2% (bls.gov) The employment, or size, of computer hardware engineers in 2020 was 66,200.

★ The projected demand for software engineers from 2020 to 2030 is 22%, while the projected demand for computer hardware engineers for the same period is 2%. Individuals having a bachelor's degree or higher will probably get hold of remarkable career opportunities, particularly if they are up-to-date with new technology advancements.

◄ EXAMPLES OF COMPUTER ENGINEERING

SAGE

In 1957 IBM began construction of its $67Billion computer named SAGE (Semi- Automatic Ground Environment). It was a significant contribution to the development of the internet. IBM built SAGE as a means to strengthen the radar and missile air defenses of the US during the Cold War. SAGE comprised of 20 connected centers: each being a one-acre concrete windowless structure. Each CPU weighed about 250 tons and could execute about 75,000 instructions per second.

TIANHE-2

Supercomputer Tianhe-2 was developed by the National University of Defense Technology (NUDT) in Changsha, China. It was deemed the number one supercomputer multiple times in the last decade for its ability to compute quadrillions (a number equal to 1 followed by 15 zeros) of calculations per second.

SOCIAL MEDIA

Social Media was birthed out of the need for online communication. After the emergence of the internet in the 1970's, there were many attempts to solve this problem. Some of the more memorable platforms are Friendster, My Space, Hi5, LinkedIn, Facebook (now Meta), Twitter, Instagram, and Tik Tok. Active social media platforms now cater to an estimated 5 billion

mobile user base. The platforms allow people to interact with others all over the globe, as well as create and share content. This is a huge accomplishment in global interaction and communication.

Soil Mechanics or Geotechnical Engineering was founded by Karl Terzaghi in 1925?

107

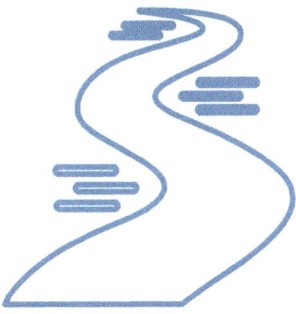

GEOTECHNICAL ENGINEERING

DESCRIPTION OF FIELD

This is another sub-specialty of civil engineering. It is the study of soil behavior under the influence of soil-water interactions and loading forces. It involves systematically applying techniques that allow construction in, on, or with geomaterials like rock and soil.

WHAT DO THEY DO?

Geotechnical engineering involves the study and engineering evaluation of earth materials such as soil, rock, and groundwater. It is a determination of the properties of unseen and variable materials that are applied to the design of structural foundations, retaining walls, earth dams, and other civil engineering works.

The goals of geotechnical engineers can include temporary excavation support, the design of foundations, and route selection for highways and railways. A geotechnical engineer engages in laboratory and field investigations for the determination of engineering properties of site soils and other geomaterials, along with their subsequent usage in the analytical study of an issue at hand.

WHERE DO THEY WORK?

Employment opportunities in this field are in governmental agencies, public utilities, specialized contractors, resource industry organizations, environmental agencies, and geotechnical and engineering consultants.

NOTABLE HISTORICAL PERSONS IN THIS FIELD

✦ Karl Terzaghi (1883-1963) was an Austrian geotechnical engineer. He is regarded as the "father of soil mechanics and geotechnical engineering" and was the first individual to elaborate a comprehensive mechanics of soils with the publication in 1925.

✦ Arthur Casagrande (1902 –1981) was an American geotechnical engineer widely known for his fundamental research on soil liquefaction and seepage and ingenious designs of soil testing apparatus.

DID YOU KNOW?

The Leaning Tower of Pisa was finally stabilized in 1993 by applying 600 tons of lead ingots as a counterweight to the north side?

◀ EXAMPLES OF GEOTECHNICAL ENGINEERING

GEOSTATIONARY OPERATIONAL ENVIRONMENTAL SATELLITES

The Geostationary Operational Environmental Satellite (GOES) assists the National Oceanic and Atmospheric Administration (NOAA) in predicting and observing weather patterns and events. The system kicked into high gear with the launch of its second generation, the GOES I-M series, which took observation times of the Earth from 10% to 100%. Launched from 1994 to 2001 and since decommissioned, GOES 9-12 unraveled the mysteries of clouds and fog, ocean currents, storms, winds, and even snow melt.

THE PANAMA CANAL

The Panama Canal is approximately 80 kilometers (about half the distance from Washington, D.C. to New York City) in length. It was constructed from 1914 through 1979. Between 2009 and 2015 the Canal was expanded to double its original capacity, to a tune of $5.2 Billion. The expansion allows new Panamax mega-ships safe passage. This feat was accomplished by deepening the Atlantic and Pacific channels as well as deepening the lake. The new locks are 55 meters wide and 427 meters (about the height of the Empire State Building) long. For a ship to pass through the three locks, 50 million gallons of fresh water are released into the sea. The modern design was done with reserve basins which retain 60% of the water.

GOLDEN GATE BRIDGE (SAN FRANCISCO, CALIFORNIA)

The Golden Gate bridge was completed in 1937. At the time, the $27 million project was said to be the tallest and longest suspension bridge on record. The construction proved to be an engineering wonder as workers had to contend with underwater blasting of rocks to install an earthquake-proof foundation. The bridge's length is 2737 meters while its longest span is over 1280 meters (about the length of 15 city blocks), suspended from two cables 227 meters high (about twice the height of the Statue of Liberty).

NUCLEAR ENGINEERING

DESCRIPTION OF FIELD

Field of engineers that research and develop the processes, instruments, and systems used to derive benefits from nuclear energy and radiation. Many of these engineers find industrial and medical uses for radioactive materials—for example, in equipment used in medical diagnosis and treatment. Many others specialize in the development of nuclear power references for ships or spacecraft.

DID YOU KNOW?

While producing electricity, nuclear power plants do not emit greenhouse gasses?

DID YOU KNOW ?

The United States was the largest producer of nuclear power in 2020?

WHAT DO THEY DO?

Nuclear engineers design nuclear equipment, monitor nuclear facility operations, examine nuclear accidents and take corrective actions in emergencies. They also use nuclear material for medical imaging devices such as PET scanners.

WHERE DO THEY WORK?

Career includes employment for for electric power organizations that utilize nuclear power plants or help maintain and service them. Others choose to work in industries that utilize radiation or radioactivity, like agriculture, medicine, and food. These fields require nuclear engineers to monitor processes, protect the public and detect problems. Additionally, the federal government hires nuclear engineers to develop advanced technologies which will be used in future power plants, design next-generation reactors for space probes, submarines, aircraft carriers, and regulate radiation or nuclear power uses. Listed below are some of the industries where nuclear engineers go on to work:

➡ Agriculture

➡ Space

➡ Energy

➡ Government

➡ Medicine

NOTABLE HISTORICAL PERSONS IN THIS FIELD

✦ **Lefteris Tsoukalasis** is regarded as one of the world's five best nuclear engineers. Invited as a Research Fellow at the Japan Atomic Energy Research Institute (Tokai-mura, Ibaraki Prefecture, Japan) for a year in 1992. Here, he was involved in researching intelligent monitoring systems for nuclear applications. He received the Humboldt Prize in 2009, which is a great honor in the field of science in Germany.

ESTEEMED PERSONS OF MINORITY

✦ **Ciara Sivels** s named the first black woman to receive a PhD in nuclear engineering from the University of Michigan. Her past research focused on treaty verification and nuclear explosion monitoring. This led to numerous research awards and four first-authored publications. She has been credited with the founding of the Women in Nuclear Engineering and Radiological Sciences (NERS) group in Michigan. This group is a space for the department's women to share stories, discuss issues, and commune.

WOMEN IN THE FIELD

✦ **Anna Biela** received a nuclear engineering degree from Purdue University. She has studied nuclear science from the time she was in grade school. Notably, her grandfather was involved in developing the world's first atomic bomb – the Manhattan Project. At the university, she distinguished herself as an ambassador for the department of nuclear engineering and often partnered with the Women in Engineering group to support female students and promote community outreach events.

DID YOU KNOW?

In 2020 nuclear energy provided America with 52% of its clean energy?

CURRENT PERCENTAGE OF WOMAN & MINORITY EMPLOYED IN THE FIELD

➡ 12.7% WOMEN

➡ 68.6% WHITE

➡ 9.1% HISPANICE

➡ 15.6% ASIAN

PROJECTED PERCENTAGE NEED FOR THE FUTURE

The projected percent change in employment from 2020 to 2030 is -8% (bls.gov) The employment, or size, of this occupation in 2020 was 17,200.

★ The projected numeric change in employment from 2020 to 2030 is -1,500 annually.

◄— EXAMPLES OF NUCLEAR ENGINEERING

NUCLEAR WEAPON

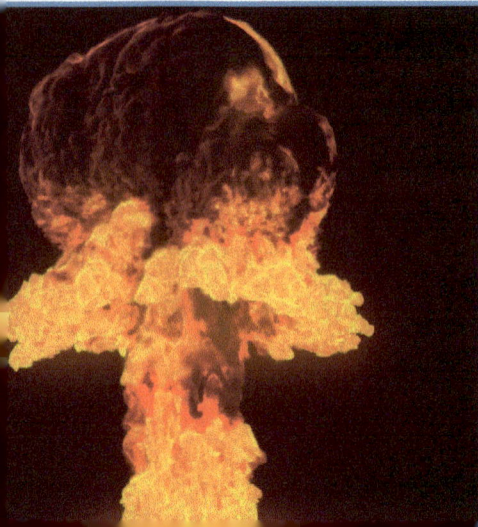

A nuclear weapon is an explosive device and is a weapon of mass destruction. It can be Fission bombs or thermonuclear bombs. The energy of nuclear weapons' yields are measured in kilotons and megatons. Quite a considerable number of such weapons remain in the world today.

The TSAR Bomba is the largest bomb ever to be detonated. It had a 100-megaton capacity and was then modified to yield 50 megatons, (which is calculated to be 3800 times the strength of the Hiroshima bomb in World War II!)

NUCLEAR REACTOR

A nuclear reactor is used to control nuclear fissions. This is the process of atomic nucleus splits, during which enormous amounts of energy are released. The energy released heats water to make steam. The steam then powers turbines which generate electricity. The conventional types of nuclear reactors are pressurized water reactors (most common), boiling water reactors, advanced gas-cooled reactors, and light water graphite reactors.

NUCLEAR SUBMARINE

A nuclear submarine is a submarine powered by a nuclear reactor, thereby converting water into steam energy to turn the propulsion turbines. They have been around from 1958 and can travel at speeds of up to 40 knots (46 mph). The weight of these submarines is measured in displacement tonnage. Nuclear powered attack submarines can be between 2,600-13,800 displacement tones. There are currently more than 15 in service.

PETROLEUM, MINING AND GEOLOGICAL ENGINEERING

DESCRIPTION OF FIELD

Mining engineers design and develop methods and strategies for extracting oil and gas from beneath the Earth's surface. These engineers also find innovative ways to extract oil and gas from older wells. Geological engineering is a specialty of geology. Both mining and geological engineers are responsible for designing mines to safely and efficiently extract coal, metals, and other substances. These are used in manufacturing and utilities.

DID YOU KNOW?

The first oil well away from land was built in 1947?

WHAT DO THEY DO?

Petroleum, geological and mining engineers design safe and efficient excavation of minerals like metals and coal for use in utilities and manufacturing. Some of the things that they typically do are:

- ✦ Offer solutions to issues related to sustainability, land reclamation, air, and water pollution
- ✦ Ensure that the operation of mines takes place in an unharmed manner
- ✦ Design underground and open-pit mines
- ✦ Prepare technical reports for managers, miners, and engineers
- ✦ Supervise the construction of tunnels and mine shafts
- ✦ Devise ways for the transportation of minerals to processing plants

Petroleum engineers develop and design ways to extract gas and oil from underground deposits. Also, they look for new methods for the extraction of gas and oil from older wells.

WHERE DO THEY WORK?

Careers include work in oil and gas extraction, management companies, mining support agencies, as well as petroleum and coal products manufacturing.
Mining and geological engineers work in other engineering firms, coal mining, metal ore mining and government.

Did you know that in 2010, a mining incident in Chile resulted in people being underground for 69 days? This is the longest time survived underground!

NOTABLE HISTORICAL PERSONS IN THIS FIELD

✦ **John Hays Hammond** *(1855 - 1936)* was a U.S. mining engineer who helped in the development gold mining in California and South Africa. In 1880, Hammond was involved in a study of the California goldfields by the U.S. Geological Survey. Then, he visited most of the countries of South and North America as a consulting engineer.

ESTEEMED PERSONS OF MINORITY

✦ **Akwasi Boakye** *(1827-1904)* from Ghana is the first black mining engineer globally. He was a prince of the Asante Empire, who trained at Delft University. Boakye was a member of the Association of Civil Engineers, which was later known as the Association of Delft Engineers. In 1871, he became a correspondent and member of the Dutch East Indies.

DID YOU KNOW?

Did you know that the first metals to be excavated were gold and copper?

WOMEN IN THE FIELD

✦ **Linda Zarda Cook** *(born in 1958)* is an American businesswoman who became the CEO of Shell Gas & Power. She graduated with a degree in petroleum engineering from the University of Kansas in 1980 and began her Shell career that same year as a reservoir engineer. Then she advanced through the organization to accept greater responsibilities in production and exploration with Shell U.S.A. In October 2002, Fortune magazine named Cook the world's 11th most powerful businesswoman. Also, Forbes magazine named her the world's 44th most powerful woman in September 2007.

CURRENT PERCENTAGE OF WOMAN & MINORITY EMPLOYED IN THE FIELD

➡ 88.1% MALE

➡ 11.2% FEMALE

➡ 73.4% WHITE

➡ 10.8% ASIAN

➡ 3.7% BLACK

MALE AVERAGE SALARY

➡ $150,061

FEMALE AVERAGE SALARY

➡ $154,718

DID YOU KNOW ?

Did you know that before Seismic Imaging became possible, oil was only found in places where it simmered to the surface?

The projected percent change in employment of petroleum engineers from 2020 to 2030 is 8%. The employment, or size, of this occupation in 2020 was 28,500.

★ The projected numeric change in employment from 2020 to 2030 is 2,200 annually.

★ The projected percent change in employment of mining and geological engineers from 2020 to 2030 is 4%. The employment, or size, of this occupation in 2020 was 6,300.

★ The projected numeric change in employment from 2020 to 2030 is 200 annually.

EXAMPLES OF PETROLEUM, MINING AND GEOLOGICAL ENGINEERING

HYDRAULIC FRACTURING

Hydraulic Fracturing is a well stimulation technique used to remove natural gas, oil, and shale from rocks. Water and sand along with other chemicals are blasted into sedimentary rocks to crack and release the trapped oil and gas. According to the American Petroleum Institute, in the United States alone, hydraulic fracturing has helped pump an extra 7 billion barrels of oil from the ground.

Did you know that Hydraulic Fracking is used to extract trapped oil from rocks?

SEISMIC IMAGING

Seismic imaging is an exploration method using high-resolution refraction imaging to map out underground geology. It is used in oil and other environmental exploration. It allows better guidance and removes the guesswork in the process of drilling for oil and other precious commodities. Coupled with supercomputers, more accurate analysis is now available to predict and understand subsurface conditions.

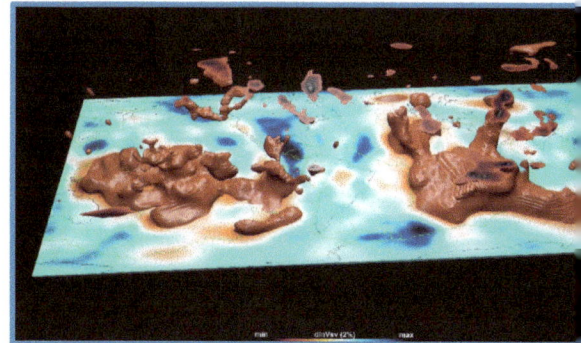

Did you know that the Big Hole of Kimberly in South Africa was dug by hand over 52 years?

HORIZONTAL DRILLING

Horizontal drilling is a method that is non-vertical. This method reduces the need to drill multiple holes in close proximity, and can reach more target areas while expanding payloads. The drill starts off in a vertical direction, and then turns in a non-vertical direction. This greatly assists in reducing the environmental footprint.

HEALTH AND SAFETY ENGINEERING

DESCRIPTION OF FIELD

Field of engineering that develop procedures and design systems to protect people from illness and injury and property from damage. They combine knowledge of engineering and of health and safety to make sure that chemicals, machinery, software, furniture, and other products will not cause harm to people or damage to property."

WHAT DO THEY DO?

Health and Safety Engineers may investigate industrial accidents, as a way to identify the cause and prevent future accidents. Fire prevention and protection engineers, product safety engineers and safety systems engineers fall under this specialty.

Health and safety engineers design systems and develop processes for the protection of individuals from injury and illness. They also work to protect property from damage. These individuals combine knowledge of health and safety along with engineering to ensure that software, furniture, chemicals, machinery, and other products will not lead to property damage and harm to any individual.

Did you know that technology innovation has significantly improved health and safety?

WHERE DO THEY WORK?

Careers include manufacturing, government and construction industries. The specific of work for a health and safety engineer depends on their background. These individuals may be employed to work for the development of health and safety equipment, chemicals, and technology. While most Health and Safety Engineers work in manufacturing, they have varied roles. They can work in a factory, on-site, or in research and development. Additionally, they may work to acquire equipment, oversee installation, and maintain it.

NOTABLE HISTORICAL PERSONS IN THIS FIELD

✦ **C. Christopher Patton,** CSP, is the president of the ASSE (American Society of Safety Engineers). He is also the principal safety, health and environmental engineer at Covidien and was awarded the prestigious ASSE Charles V. Culbertson Outstanding Volunteer Service award for 2006-07.

CURRENT PERCENTAGE OF WOMAN & MINORITY EMPLOYED IN THE FIELD

➡ 18.4% WOMEN

➡ 71.1% WHITE

➡ 10.7% LATINO or HISPANIC

➡ 11.6% ASIAN

PROJECTED PERCENTAGE NEED FOR THE FUTURE

The projected percent change in employment from 2020 to 2030 is 6% (bls.gov) The employment, or size, of this occupation in 2020 was 24,100.

★ The projected numeric change in employment from 2020 to 2030 is 1,500 annually.

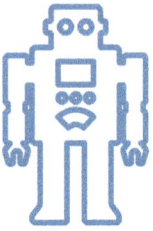

INDUSTRIAL ENGINEERING

DESCRIPTION OF FIELD

These engineers are tasked with coming up with plans and strategies to eliminate wastefulness in production processes. These engineers develop systems that are efficient and reliable to integrate information, materials, machines, workers, and energy to make a product or provide a service.

Their skills can be applied to a variety of settings including healthcare systems and business administration. Some focus entirely on the automated aspects of the manufacturing process and are called manufacturing engineers.

WHAT DO THEY DO?

Careers options for industrial engineers include work in every stage of processing and production. They design new facilities from the ground up, and expand, reconfigure or upgrade existing facilities. Additionally, they write specifications for equipment bought from outside vendors or design new equipment. Furthermore, they design new processes, new tools, and fixtures, and repurpose existing equipment and facilities.

WHERE DO THEY WORK?

Industrial Engineers work in transportation equipment manufacturing, computer, and electrical product manufacturing as well as scientific and technical services. They work in several different environments ranging from offices to various settings that they are tasked with improving. Their roles can include examining workflows in a hospital or watching the way a process works in a factory.

These engineers feature a wide array of skills, so they can work in both technical and managerial positions. Also, they can be found in different employment settings - research and development, consulting and engineering, manufacturing and trade, service industries, and logistics.

NOTABLE HISTORICAL PERSONS IN THIS FIELD

✦ **Henry Ford** (1863-1947) is perhaps the most renowned industrial engineer. He installed the first moving assembly line for the mass production of a complete automobile. Ford's innovation reduced the time taken to construct a car from over 12 hours to 2 hours and 30 minutes.

DID YOU KNOW?

Did you know that an industrial engineer can assist in reducing the waiting times at an amusement park?

WOMEN IN THE FIELD

✦ **Lillian Gilbreth** *(1878-1972)* is known as the "mother of modern management." She pioneered industrial management techniques that are used today. She was a consultant to several industrial firms, including General Electric. Her work in these firms improved the design of household and kitchen appliances. Some of the innovative techniques that she designed were aimed at specifically helping disabled women in completing common household tasks.

CURRENT PERCENTAGE OF WOMAN & MINORITY EMPLOYED IN THE FIELD

➡ 19.8% WOMEN

➡ 68.1% WHITE

➡ 10.6% LATINO or HISPANIC

➡ 14.8% ASIAN

PROJECTED PERCENTAGE NEED FOR THE FUTURE

The projected percent change in employment from 2020 to 2030 is 14% (bls.gov)

The employment, or size, of this occupation in 2020 was 292,000.

★ The projected numeric change in employment from 2020 to 2030 is 40,000 annually.

EXAMPLES OF INDUSTRIAL ENGINEERING

INDUSTRIAL ROBOTICS

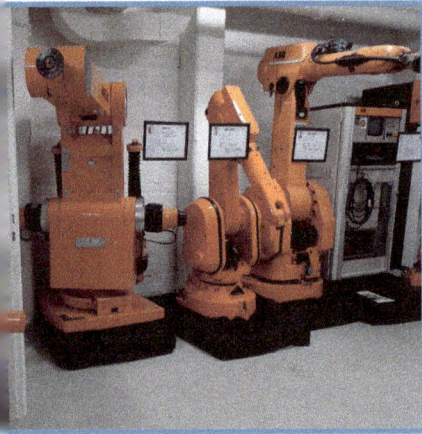

The first industrial arm made its debut at General Motors in 1961, performing tasks too dangerous for workers. In 2006 NASA and General Motors signed the Space Act Agreement for the construction of Robonaut 2. After completion, they recognized commonalities that would assist auto workers with fatigue and better grip and thus aimed to build the first industrial strength glove called Robo-Glove.

AUTOMATED WAREHOUSES AND SMART RECYCLING

Ocado Technologies has built an online warehouse powered by Artificial Intelligence (AI) and robotics. Think Amazon, but bigger! Ocado is the biggest online grocery retail store where the robots move at faster speeds — approximately 13 feet per second. Kroger, a US grocery chain has teamed up with Ocado to bring this UK model to the United States.

DID YOU KNOW?

Did you know that Industrial Engineering focuses on both people and technology?

GLOSSARY

ACADEMIA - the life, community, or world of teachers, schools, and education.

ACCREDITED - the state in which an educational institution has been recognized as maintaining standards that qualify the graduates for admission to higher or more specialized institutions or for professional practice

AEROPONICS - the process of growing plants without the use of soil by suspending their roots in the air and spraying them with nutrient-rich mist.

ASTRONAUT - A spaceman or spacewoman, especially one who has been trained and certified by the National Aeronautics and Space Administration (NASA), European Space Agency (ESA), Canadian Space Agency (CSA), or Japan Aerospace Exploration Agency (JAXA).

ASTRONOMY - the study of matter and objects outside the earth's atmosphere such as their physical and chemical properties.

CALCULUS - a method of computation or calculation comprising differential and integral calculus in a special notation

CAPSTONE - this exam is used to assess one's cumulative knowledge over a period of study.

COMPRESSORS - a machine that supplies air or other gases at an increased pressure.

COSMONAUT - Λ person trained and certified by the Russian Space Agency to work in space.

ELECTRICAL - that which is related to, or operated by electricity

ELECTROMAGNETISM - this branch of physical science is concerned with the physical relations between electricity and magnetism. It is responsible for interactions between charged particles.

HYDRAULICS - a branch of science that deals with practical applications (such as the transmission of energy or the effects of flow) of liquid (such as water) in motion.

HYDROELECTRIC - of or relating to the production of electricity by water power.

HYDROPONICS - the growing of plants in nutrient solutions with or without an inert medium (such as soil) to provide mechanical support.

KINEMATICS - a branch of dynamics that deals with aspects of motion apart from considerations of mass and force

LUCRATIVE - producing wealth; profitable.

MECHANICAL - of or related to machinery or manual operations.

PROSTHETICS - he surgical or dental specialty concerned with the design, construction, and fitting of prostheses.

REACTOR -

i) a device (such as a coil, winding, or conductor of small resistance) used to introduce reactance into an alternating-current circuit

ii) a device for the controlled release of nuclear energy (as for producing heat)

REALM - refers to a sphere or domain

SIEGE - a military blockade of a city or fortified place to compel it to surrender

SPECIALTIES - something in which one specializes

STATISTICS - a branch of mathematics dealing with the collection, analysis, interpretation, and presentation of masses of numerical data

STRUCTURAL - of, relating to or affecting structures

SURVEYING - a branch of applied mathematics that deals with the collection, analysis, interpretation, and presentation of masses of numerical data determining the area of any portion of the earth's surface, the lengths and directions of the bounding lines, and the contour of the surface. In Engineering we focus on Geodetic Survey and Plane Surveying.

GEODETIC SURVEYING - a type of surveying that determines the exact location of permanent points on the surface of the earth, factoring in its shape, size and curvature.

PLANE SURVEYING - ordinary field and topographical surveying in which the curvature of the earth is not considered.

UNDERGRADUATE - refers to a student of a college or university who does not have a first (especially Bachelor's) degree.

VENTILATION - the circulation of air

VOLTAGE - electric potential or potential difference expressed in volts

ABOUT THE AUTHOR

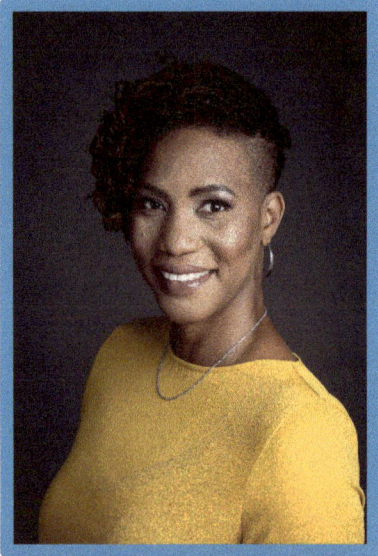

MARCIA ROBIN-STOUTE is a dedicated educator with over 15 years of experience, specializing in engineering instruction. She's passionate about teaching and has guided university students from diverse backgrounds in courses like Transportation Engineering, Civil Materials Lab, Geotechnical Engineering Lab, and Alternative Energy. Marcia also excels as a mentor, assisting university students in selecting majors, internships placements and leading student organizations.

In addition to the classroom, Marcia manages summer Pre-College engineering programs for underrepresented minority students in STEM fields, earning recognition from funding partners, students, and parents.

Beyond her academic pursuits, Marcia is a devoted wife and mother of three. She believes in giving back to those who've supported her journey and is determined to be a voice for underrepresented individuals in STEM. Her ultimate goal? To see more diversity in the STEM fields, especially among women and people of color.

REFERENCES

CAREER & LABOR STATISTICS
www.archihan.tripod.com
www.bls.gov
www.datausa.io
www.indeed.com
www.loc.gov
www.twi-global.com
www.zippia.com

STUDENT CAREER RESOURCES
www.academicinfluence.com
www.admissions.psu.edu
www.careerexplorer.com
www.collegedunia.com
www.degreequery.com
www.energy.gov
www.futuresinengineering.org
www.indeed.com
www.internationalstudent.com
www.iveyengineering.com
www.learn.org

www.macfound.org
www.owlcation.com
www.raise.me
www.thehistorymakers.org

ENGINEERING (CAREER) RESOURCES
www.britannica.com
www.engr.ncsu.edu
www.engineerdaily.com
www.futuresinengineering.com
www.nspe.org
www.ppi2pass.com
www.theweldinginstitute.com

NOTABLE PERSONS IN ENGINEERING
www.aaregistry.org
www.aimehq.org
www.aerielviews.blog
www.blackenterprise.com
www.blackhistory.com
www.blackpast.org
www.electricalestimates.co.uk
www.ellines.com
www.energy.gov
www.engineersdaily.com
www.face2faceafrica.com
www.geoengineer.org
www.ghanaianmuseum.com
www.gizmodo.com
www.goldieblox.com
www.greedhead.net
www.jacobsinstitute.berkeley.edu

www.mccormick.northwestern.ed
www.michigandaily.com
www.nae.edu
www.nkaa.uky.edu
www.newsamericasnow.com
www.peoplepill.com
www.ranker.com
www.thefamouspeople.com
www.web.iit.edu
www.workflowmax.com
www.veranda.com
www.zackgh.com

ENGINEERING SPECIALIZATIONS:
AGRICULTURE
www.agriculture.mo.gov
www.foodtank.com

BIOMEDICAL
www.uchicagomedicine.org

CIVIL
www.asce.org
www.ice.org.uk
www.mcgill.ca

COMPUTER
www.computerscience.org

ELECTRICAL
www.beckhearingaids.com
www.electricalestimates.co.uk

ENVIRONMENTAL
www.be3corp.com
www.creativesafetysupply.com
www.environmentalscience.org
www.livescience.com
www.nationalgeographic.org
www.oceanpowertechnologies.com

GEOTECHNICAL
www.geoengineer.org
www.geoengineeringwatch.org
www.sciencedirect.com

MECHANICAL
www.engineerine.com
www.towerfast.com
www.wylerindustrial.com

NUCLEAR
www.nuce.psu.edu

PETROLEUM, MINING & GEOLOGICAL
www.aimehq.org
www.rigzone.com

STRUCTURAL
www.ae.psu.edu
www.marineinsight.com
www.seaonc.org
www.structuralengineeringbasics.com
www.theweldinginstitute.com